India-Japan Expert Group Meeting on Biomolecular Electronics & Organic Nanotechnology for Environment Preservation (IJEGMBE 2015)

Journal of Physics: Conference Series Volume 704

Fukuoka, Japan
23 – 26 December 2015

Editors:

Akihiko Fujii
Bansi D. Malhotra

ISBN: 978-1-5108-2285-6
ISSN: 1742-6588

Printed from e-media with permission by:

Curran Associates, Inc.
57 Morehouse Lane
Red Hook, NY 12571

Some format issues inherent in the e-media version may also appear in this print version.

Printed by Curran Associates, Inc. (2016)

For permission requests, please contact the Institute of Physics
at the address below.

Institute of Physics
Dirac House, Temple Back
Bristol BS1 6BE UK

Phone: 44 1 17 929 7481
Fax: 44 1 17 920 0979

techtracking@iop.org

Additional copies of this publication are available from:

Curran Associates, Inc.
57 Morehouse Lane
Red Hook, NY 12571 USA
Phone: 845-758-0400
Fax: 845-758-2633
Email: curran@proceedings.com
Web: www.proceedings.com

Table of contents

Volume 704

India-Japan Expert Group Meeting on Biomolecular Electronics & Organic Nanotechnology for Environment Preservation (IJEGMBE 2015)23–26 December 2015, Fukuoka, Japan

Accepted papers received: 23 March 2016Published online: 19 April 2016Preface

011001
OPEN ACCESS
Foreword

011002
OPEN ACCESS
Peer review statement

Papers

Organic iontronics

012001
OPEN ACCESS
A study on the stability of n-type conductive polymer

M Onoda pg. 1

012002
OPEN ACCESS
Parametric Optimization of Experimental Conditions for Dye-Sensitized Solar Cells based on Far-red Sensitive Squaraine Dye

Takuya Morimoto, Naotaka Fujikawa, Yuhei Ogomi, Shyam S. Pandey and Shuzi Hayase pg. 13

012003
OPEN ACCESS
Tandem Dye-Sensitized Solar Cells Based on TCO-less Back Contact Bottom Electrodes

Ajay K. Baranwal, Naotaka Fujikawa, Terumi Nishimura, Yuhei Ogomi, Shyam S. Pandey, Tingli Ma and Shuzi Hayase pg. 20

012004
OPEN ACCESS
Research Trends of Soft Actuators based on Electroactive Polymers and Conducting Polymers

K Kaneto pg. 28

Fabrication and evaluation

012005
OPEN ACCESS
Orientation Characteristics of Non-regiocontrolled Poly (3-hexyl-thiophene) Film by FTM on Various Liquid Substrates

M Pandey, S Nagamatsu, S S Pandey, S Hayase and W Takashima pg. 37

012006
OPEN ACCESS
Ambipolar Carrier Mobility in Binary Blend Thin Film of Non-Peripheral Alkylphthalocyanines

Akihiko Fujii, Shohei Nakano, Hitoshi Fukui, Takashi Saito, Masashi Ohmori, Yo Shimizu and Masanori Ozaki pg. 44

012007
OPEN ACCESS
Electrophoretically deposited multiwalled carbon nanotube based amperometric genosensor for *E.coli* detection

Hema Bhardwaj, Shipra Solanki and Gajjala Sumana pg. 51

012008
OPEN ACCESS
Polaron states and charge dynamics in doped conjugated polymers studied by various spectroscopic methods

Hideo Kishida pg. 60

012009
OPEN ACCESS
Amphiphilic Fluorinated Polymer Nanoparticle Film Formation and Dissolved Oxygen Sensing Application

Yu Gao, Huie Zhu, Shunsuke Yamamoto, Tokuji Miyashita and Masaya Mitsuishi pg. 65

Molecular electronic materials, including bio-sensing materials

012010
OPEN ACCESS
Conducting paper based sensor for cancer biomarker detection

Saurabh Kumar, Suveen Kumar, Chandra Mouli Pandey and Bansi D. Malhotra pg. 73

012011
OPEN ACCESS
Nanomaterials based biosensors for cancer biomarker detection

Bansi D Malhotra, Saurabh Kumar and Chandra Mouli Pandey pg. 82

012012
OPEN ACCESS
Photophysical investigations of squaraine and cyanine dyes and their interaction with bovine serum albumin

M Saikiran, D Sato, S S Pandey and T Kato pg. 93

012013
OPEN ACCESS
Functionalized vertically aligned ZnO nanorods for application in electrolyte-insulator-semiconductor based pH sensors and label-free immuno-sensors

Narendra Kumar, Sujata Senapati, Satyendra Kumar, Jitendra Kumar and Siddhartha Panda pg. 103

012014
OPEN ACCESS
Antibody conjugated graphene nanocomposites for pathogen detection

Chandan Sign and Gajjala Sumana pg. 111

012015
OPEN ACCESS
DNA methylation detection based on difference of base content

Shinobu Sato, Keiichi Ohtsuka, Satoshi Honda, Yusuke Sato and Shigeori Takenaka pg. 118

Electronic devices for printed electronics

012016
OPEN ACCESS
Influence of emissive layer thickness on electrical characteristics of polyfluorene copolymer based polymer light emitting diodes

D Das, P Gopikrishna, A Singh, A Dey and P K Iyer pg. 126

012017
OPEN ACCESS
High Performance, Low Operating Voltage n-Type Organic Field Effect Transistor Based on Inorganic-Organic Bilayer Dielectric System

A Dey, A Singh, A Kalita, D Das and P K Iyer pg. 131

012018
OPEN ACCESS
Improved characteristics of polymer light-emitting diode based on poly(alkylfluorene) with functional layer fabricated by contact printing method

H Kajii, K Mizuno and Y Ohmori pg. 139

Environmental preservation

012019
OPEN ACCESS
Growth promotion effect of steelmaking slag on *Spirulina platensis*

R Nogami, L T Tam, H T L Anh, H T H Quynh, L T Thom, P V Nhat, N T H Thu, D D Hong and M Wakisaka pg. 143

012020
OPEN ACCESS
Cascaded H-bridge multilevel inverter for renewable energy generation

Ravikant Pandey, Ravi Nath Tripathi and Tsuyoshi Hanamoto pg. 151

Liquid crystal

012021
OPEN ACCESS
Analysis of Dynamic Response of Permittivity in a Liquid Crystal Cell with Flow

R Ozaki, K Matsuura, K Kadowaki, T Q Duong and H Moritake pg. 157

012022
OPEN ACCESS
Supramolecular Nanocomposites: Dispersion of Zero-, One- and Two-dimensional Nanoparticles in Discotic Liquid Crystals

Sandeep Kumar pg. 162

Journal of Physics: Conference Series **704** (2016) 011001 doi:10.1088/1742-6596/704/1/011001

December 22-25, 2015
Nakamura Centenary Memorial Hall
Kyushu Institute of Technology

India-Japan Expert Group Meeting on Biomolecular Electronics & Organic Nanotechnology for Environment Preservation
(IJEGMBE)

FOREWORD

There is increased interest in organic nanotechnology and biomolecular electronics for environmental preservation, and in their anticipated impact on the economics of both the developing and the developed world. Keeping this in mind, the Department of Biological Functions, Graduate School of Life Sciences and Systems Engineering, Kyushu Institute of Technology (KIT), Kitakyushu, Japan, and the Department of Science & Technology Centre on Biomolecular Electronics (DSTCBE), National Physical Laboratory (NPL) jointly organized the India-Japan Workshop on Biomolecular Electronics and Organic Nanotechnology for Environmental Preservation (IJWBME 2009) at NPL, New Delhi from 17th – 19th December 2009, IJWBME 2011 at EGRET Himeji, Himeji, from 7th – 10th December, Japan, and IJWBME 2013 at Delhi Technological University, New Delhi, from 13th – 15th December. The India-Japan Expert Group Meeting on Biomolecular Electronics & Organic Nanotechnology for Environment Preservation (IJEGMBE) will be held from 22th – 25th, December, 2015, at Nakamura Centenary Memorial Hall, Kyushu Institute of Technology, Kitakyushu, Japan in association with Delhi Technological University, Delhi, India.

Recent years have seen rapid growth in the area of Biomolecular Electronics involving the association and expertise of physicists, biologists, chemists, electronics engineers and information technologists. There is increasing interest in the development of nanotechnology and biomolecular electronic devices for the preservation of our precious environment. In this context, the world of the electronics, which developed on Si semiconductors, is going to change drastically. A paradigm shift towards organic or printed electronics is more likely in the future. The field of organic electronics promises exciting new technologies based on inexpensive and mechanically flexible electronic devices, and is now starting to see commercial success. On the sidelines of this increasingly well-established field are several emerging technologies with innovative mechanisms and functions that utilize the mixed ionic/electronic conducting character of conjugated organic materials. These techniques are based around flexible or printed electronics. Ionic functionalization influences many of the key properties of conductive polymers through its impacts on molecular order, the injection and transport of charge, optical excitations, and interactions with other molecules. Consequently, it is an important tool in the development of electronic and photonic devices based on conductive polymers. We have considered that to focus exclusively on the iontronics, ionic carriers in organic electronic materials and devices in organic electronic materials seems timely. Therefore, this report reviews the scientific understanding and important scientific discoveries made in the electrochemistry of conductive polymers based on our experience.

The main purpose of ***IJEGMBE*** is to provide an opportunity for researchers, who are interested in biomolecular electronics and organic nanotechnology for environmental preservation, to come together in an informal and friendly atmosphere and exchange their technical information and experience. We are sure that this meeting will be very useful and fruitful for all participants to summarize the recent progress in biomolecular electronics and organic nanotechnology for environmental preservation and prepare the next step for future generations.

Many papers have been submitted from India and Japan and **more than 20 papers** have been accepted for presentation. All the papers accepted will be presented.

The main topics of interest are as follows;

- Bioelectronics
- Biomolecular Electronics
- Fabrication Techniques
- Self-assembled monolayers
- Nano sensors
- Environmental monitoring
- Organic devices
- Organic Functional Materials
- Others

The program of this meeting consists of Invited Lectures and oral presentations. We hope all participants benefited from this meeting. We would like to express our sincere thanks to the organizing committee members of this workshop and thank organizations such as by Japan Society for the Promotion of Science (JSPS), Japan, Department of Science & Technology (DST), Govt of India, Delhi Technological University (DTU), India, Kitakyushu City, and West Japan Industry and Trade Convention Association for their financial support. Thanks are also given to University of Hyogo and the Kyushu Institute of Technology. Many people have made efforts so as to make this meeting possible and valuable, and we would like to express our thanks to Prof. Wataru Takashima, Prof. Shuichi Nagamatsu, Prof. Shyam S. Pandey, and Ms. Megumi Furuta of Kyushu Institute of Technology, Prof. Ashutosh Sharma (DST, India), Dr Navin Vasishta (DST, India), Prof. Yogesh Singh (DTU, India), Prof. Akihiko Fujii and Dr. Hirotake Kajii from Osaka University for their efforts in planning and arranging this expert group meeting.

Finally, we hope that many young and active researchers who are participating will enjoy stimulating discussions and exchanging their ideas with each other at the IJEGMBE being organized at Kitakyushu in Japan.

December 22, 2015

IJEGMBE Chair

Mitsuyoshi Onoda
Graduate School of Engineering
University of Hyogo
Himeji, Japan

IJEGMBE Chair

Bansi D. Malhotra
Department of Biotechnology
Delhi Technological University
Delhi, India

Journal of Physics: Conference Series **704** (2016) 011001 doi:10.1088/1742-6596/704/1/011001

International Advisory Committee

< Honorary Chair >
Keiichi KANETO (Osaka Institute of Technology, Japan)

< Chairs >
Mitsuyoshi ONODA (University of Hyogo, Japan)
Bansi D. MALHOTRA (Delhi Technological University, India)

< Vice-Chairs >
Shuzi HAYASE (Kyushu Institute of Technology, Japan)
Sandeep KUMAR (Raman Research Institute, Bangalore, India)

< Members >
E.S.R. GOPAL (Indian Institute of Science, Bangalore, India)
Parameswar K. IYER (Indian Institute of Technology, Guwahati, India)
S. K. JOSHI (CSIR-National Physical Laboratory, New Delhi, India)
Hideo KISHIDA (Nagoya University, Japan)
Pradeep KUMAR (Delhi Technological University, Delhi, India)
Yogesh SINGH (Delhi Technological University, Delhi, India)
Masaya MITSUISHI (Tohoku University, Japan)
Ryotaro OZAKI (Ehime University, Japan)
Milan K. SANYAL (Saha Institute of Nuclear Physics, Kolkata, India)
Ashutosh SHARMA (Department of Science & Technology, India)
Ajay K. SOOD (Indian Institute of Science, Bangalore, India)
Shigeori TAKENAKA (Kyushu Institute of Technology, Japan)
George THOMAS (Indian Institute of Science Education & Research, Kerala, India)

< Secretariat >
Wataru TAKASHIMA (Representative: Kyushu Institute of Technology, Japan)
Akihiko FUJII (Osaka University, Japan)
Hirotake KAJII (Osaka University, Japan)
Shuichi NAGAMATSU (Kyushu Institute of Technology, Japan)
Shyam S. PANDEY (Kyushu Institute of Technology, Japan)

IEME JGB 2015

***India-Japan Expert Group Meeting on Biomolecular Electronics & Organic Nanotechnology for Environment Preservation* (IJEGMBE)**
December 22-25, 2015
Nakamura Centenary Memorial Hall, Kyushu Institute of Technology

2015 IJEGMBE Attendees

IJEGMBE 2015 IOP Publishing
Journal of Physics: Conference Series **704** (2016) 011001 doi:10.1088/1742-6596/704/1/011001

Journal of Physics: Conference Series, IOP Publishing

Special Issue of IJEGMBE

Preface

This special issue contains a selection of papers presented at India-Japan Expert Group Meeting on Biomolecular Electronics & Organic Nanotechnology for Environment Preservation (IJEGMBE), which was held at the Kyushu Institute of Technology, Fukuoka, from December 22 to 25, 2015. The purpose of the meeting was to discuss the recent progress in organic nanotechnology and biomolecular electronics for environmental preservation. The total of 24 papers were presented, which included 5 invited talks and 19 oral presentations.

The 22 papers were selected for publication in the Journal of Physics: Conference Series. The selection was carried out by the editorial judgement based on the peer review process. The aim of this special issue is to circulate the information presented at the meeting to readers interested in biomolecular electronics and organic nanotechnology.

The editorial board for the special issue of the IJEGMBE would like to appreciate all the reviewers for their excellent work and efforts with the kind commitment of time and expertise. Their judgments and suggestions for the submitted papers were most valuable for maintaining the highest quality and standard for this journal. Finally, we would like to thank all authors for preparing their manuscripts and for their perseverance during the peer review process. We sincerely hope that readers appreciate the significance and distinguished potential of these fields, which will be a scientific foundation and pioneer technology in the 21st century.

March 25, 2016

Editor-in-Chief

Akihiko Fujii
Graduate School of Engineering
Osaka University
Suita, Japan

Editor-in-Chief

Bansi D. Malhotra
Department of Biotechnology
Delhi Technological University
Delhi, India

Editorial Board of IJEGMBE

Journal of Physics: Conference Series **704** (2016) 011001 doi:10.1088/1742-6596/704/1/011001

< Sponsored by >

- ***Japan Society for the Promotion of Science (JSPS), Japan***
- ***Department of Science & Technology (DST), India***

< Supported by >

- ***Kitakyushu City***
- ***West Japan Industry and Trade Convention Association***

< in cooperation with >

- **Municipal University Corporation**
  University of Hyogo
- **National University Corporation**
  Kyushu Institute of Technology

Peer review statement

All papers published in this volume of *Journal of Physics: Conference Series* have been peer reviewed through processes administered by the proceedings Editors. Reviews were conducted by expert referees to the professional and scientific standards expected of a proceedings journal published by IOP Publishing.

Journal of Physics: Conference Series **704** (2016) 012001 doi:10.1088/1742-6596/704/1/012001

A study on the stability of n-type conductive polymer

M Onoda[1]
Graduate School of Engineering, University of Hyogo
2167 Shosha, Himeji, Hyogo 671-2280, Japan

E-mail: ss-syu02@eng.u-hyogo.ac.jp

Abstract. Novel n-type conducting polymer, poly(*p*-methylpyridinium vinylene), PMePyV were synthesized by using the quaternization of poly(*p*-pyridyl vinylene), PPyV and several regiochemical consequences in this polymer were proposed. The electrical, optical, and electrochemical properties of n-type conductive polymer were observed. In addition, a possibility of a variety of functional applications of n-type conductive polymer was indicated.

1. Introduction

At present, electronic devices sweeping over a market are inorganic semiconductor devices mainly on the Si, especially Si VLSIs (Very Large-Scale Integrated circuits) which incorporated a lot of ultrafine electronic circuits. When MOS (Metal-Oxide-Semiconductor) transistors are reduced steadily and to 1/k (k is an arbitrary constant), they are packed the tip of the same area, as a result, the degree of integration and the movement speed become k^2 and k times, respectively, and the consumption electricity is controlled by $1/k^2$. Therefore, there are no factors which device performance grows to worse due to ultrafine processing. So-called, as shown in figure 1, according to scaling law established by Robert H. Dennard of IBM in 1974 [1], the Si semiconductor industry continued progressing by breaking the limit such as ultrafine processing steadily. However, it is said the in early 21st century, inorganic semiconductor electronics began to sputter due to the physical limit of the ultrafine processing technology. As the reasons, for example,

① The thickness of the gate insulation film becomes thin in the thickness of several atomic layers, as a result the restraint of the tunneling current is difficult.

② Because the density of impurities atom is too low and cannot control it, the unevenness of device properties is caused.

③ Due to various kinds of parasitism effects, the performance enhancements such as speedup based on simple scaling law, low consumption electricity, high integrations, *etc* cannot be expected.

etc.

Furthermore, development of the extreme ultraviolet ray source of light to use for lithography is an important problem.

If current electronics technologies mainly on the Si have limits, the basic solution is to break the limit by devising replacements. The suggested technique is the concept of the molecular electronic devices which is going to fabricate molecular electronics using molecules. This thought is proposed for the theoretical consideration of the diode device function consisting of one organic molecule by

[1] To whom any correspondence should be addressed.

Arieh Aviram et al. for the first time in 1974 [2], and basically are incorporating a function to control electronic flow to individual molecules and to realize electronic devices of the molecular size.

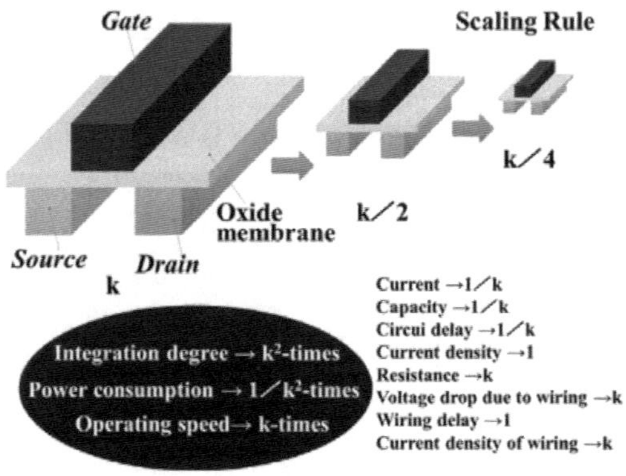

Figure 1. Image of scaling law.

On the other hand, the ability for complicated information processing based on pattern recognition or the parallel computation in brain and the nervous system of the animal is extremely high and is performed easily. The bio-system has complicated and extremely high function. That is, bioelectronics attracts great interest in making good use for converting information, taking in the function that is going to store, and also limited development of electronics technologies based on a scaling law [3].

At all events, because a variety of the organic materials have much metastable structure, the flexibility is bigger than an inorganic material markedly, and the expectations for the organic materials are particularly high.

In such a background, the world of the electronics which developed mainly on a Si semiconductor is going to change drastically. A paradigm shift which is called organic electronics or printed electronics is finally more likely.

Organic electronics based on Flexible-Organic Semiconductors are called Flexible Electronics or Plastic Electronics. In addition, because a print technology is applicable to fabricate films, the motivations to use polymers are twofold: the possibility to shrink the circuit size to the molecular level and the printing of large-area circuits at extremely low-cost. So-called this is printed electronics. The development research of organic electronic devices having active layers consisting of flexible-organic semiconductor materials such as organic light-emitting devices, organic transistors, organic solar batteries etc. is developed extremely actively [4].

In the organic semiconductors, we often call them p- or n-type materials. This is not a meaning having holes and electrons as majority carriers like an inorganic semiconductor, but the meaning that electrons and holes injection are easy. This is extremely important and should be careful [5].

As one of the most characteristic effect that conductive polymers possess, doping effects are well known. By way called the doping, if acceptor (electron-accepting) or donor (electron-donating) are soaked or put forcibly into conductive polymers, their conductivities can be widely controlled from the insulator region, via the semiconductor region, and finally to the metallic region. As a result, other characterizations such as magnetic and optical properties are also greatly changed. Basically, such doping effects are strongly caused by the change of electronic states of conductive polymers, because of the interaction between dopants and the polymer main chains. As for such conductive macromolecule, the use as the (i) metals, (ii) insulators and semiconductors, (iii) insulator/metal transition phenomena, (iv) reversible doping ability, and so on is possible [6].

Journal of Physics: Conference Series **704** (2016) 012001 doi:10.1088/1742-6596/704/1/012001

To date, almost all of the conductive polymers that have been examined for organic devices are good holes conductors and relatively poor electrons conductors. The reason why the stability of the n-type conductive polymers is remarkably low is caused by carbanion (ions of the organic compounds with a negative charge on carbon atoms, that is $-:CH_3$, here $\cdot$ indicaded unpaired electron) formed from doping easily reacting with a very small amount of oxygen and water in an atmosphere [7].

In order to fully exploit the potential offered by polymers the device materials have to be processable in its polymeric form. By appropriate modification of the molecular structure of these polymers such as the introduction of long alkyl side chains, conductive polymers which consist of rigidly conjugated double bonds have been found to be soluble in any solvents and even to be fusible at relatively low temperature.

Accordingly, the ability to fabricate organic devices from solution process promises a significant manufacturing advantage. M. J. Marsella *et al.* [8] have reported the preparation of orange-yellow poly(*p*-pyridyl vinylene), PPyV using a modification of a synthesis initially developed by R. Galarini *et al.* [9] Although PPyV can be viewed as an alternating copolymer of pyridine and acetylene and is a conductive polymer having n-type semiconductor natures, its fundamental physical properties have yet to be clarified. In addition, for example, in the presence of trifluoromethanesulfonate (CF_3SO_2OR, R: alkyl group), PPyV alkylates by heat-treatment using a suitable solvent, and poly(p-alkylpyridinium vinylene) is provided.

For example, as shown in figure 2, quaternization of the PPyV to the yellow poly(*p*-methylpyridinium vinylene), PMePyV is obtained on all isomers (Random, Head-Tail, and Head-Head forms) by reaction in neat methyl triflate in dichloromethane, CH_2Cl_2 at 40 °C for a week under argon. PMePyV has triflate ions ($-CF_3SO_3^-$) as a counter-ion [8].

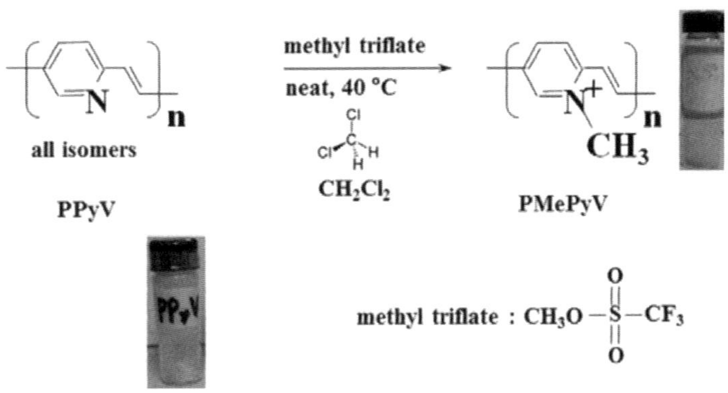

Figure 2. Conversion reaction of PPyV to yield alkylated PPyV. Here, quaternization of the PPyV to the yellow poly(*p*-methylpyridinium vinylene), PMePyV is shown.

To make the stability or conduction mechanisms of positively charged n-type conductive polymer, PMePyV clear, their fundamental physical properties such as electrical, optical and electrochemical properties have been studied and discussed from molecular structures of their reduced states [10].

2. Sample and Experimental Procedure

2.1. Regiocontrol of the PPyV structure

It is well known that while regiocontrol of polymers is an important aspect from synthetic viewpoint, the regiochemical consequences may affect physical properties of polymers [11]. Poly(3-alkylthiophene)s provide for the first time well defined structures for the investigation of structure-property relationships in this class of electronic and photonic materials. Therefore, for similar consideration, since pyridine-vinyl group of PPyV is not a symmetrical molecule, there are three

Journal of Physics: Conference Series **704** (2016) 012001 doi:10.1088/1742-6596/704/1/012001

relative orientations available when two pyridine rings are coupled between the 2- and 5-positions through a vinyl group. PMePyV has three regioisomeric forms: head-to-head (Head-Head) (Here, tail-to-tail (Tail-Tail) is the same as Head-Head), head-to-tail (Head-Tail), and Random (combination of Head-Head and Head-Tail) as shown in figure 3. Of course for similar consideration PPyV has also three regioisomeric forms.

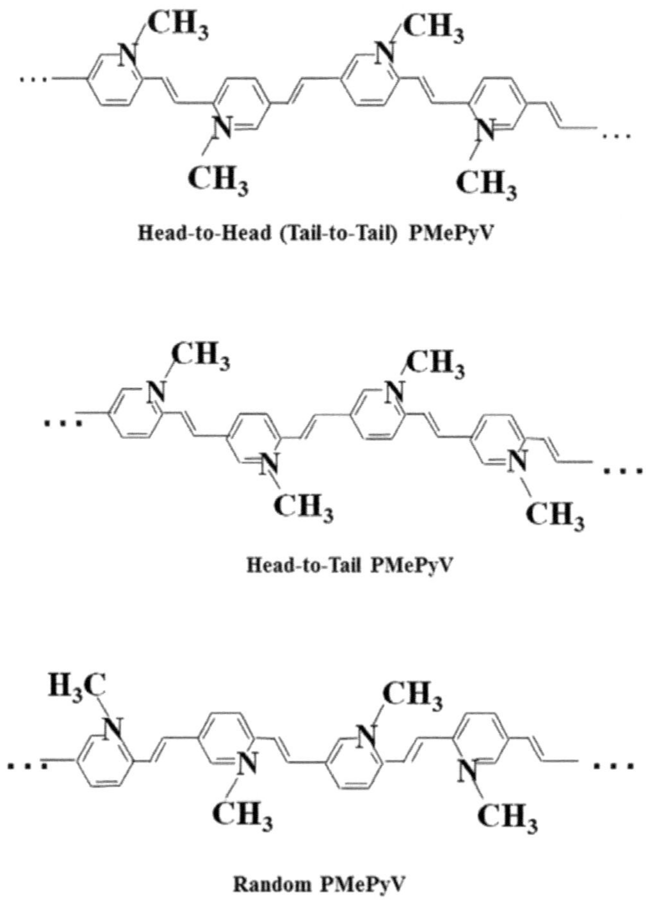

Figure 3. Possible regiochemical couplings of PMePyV.

A schematic diagram of the synthetic route to three regioisomeric forms of PPyV is outlined in figure 4. Head-Head, Head-Tail, and Random PPyVs were synthesized via extensive use of palladium catalyzed cross-coupling of vinyl stannanes and pyridine bromides. [8], [9], [12], [13] Random PPyV was prepared via cross-coupling of 2, 5-dibromopyridine and 1, 2-bis(tributylstann- yl)ethylene in the presence of palladium phosphine complex essentially according to the procedure reported by Galarini *et al.* [9] That is, 2, 5-dibromopyridine and tetrakis(triphenylphosphine)- palladium(0) were added to a solution of 1-methyl-2-pyrolidinone and the solution was stirred at 110 °C for 24 hours. The yellowish-orange precipitate was collected and washed with hot tetrahydrofuran (THF). The polymer was purified by Soxhlet extraction with ethanol for 24 hours. As shown in figure 4, the starting monomer for the desired polymer dominated the regiocontrol of the parent PPyV. For example, the synthesis of the Head-Head monomer was accomplished by the high reactivity of the 2-position of 2, 5-dibromopyridine.

Journal of Physics: Conference Series **704** (2016) 012001
doi:10.1088/1742-6596/704/1/012001

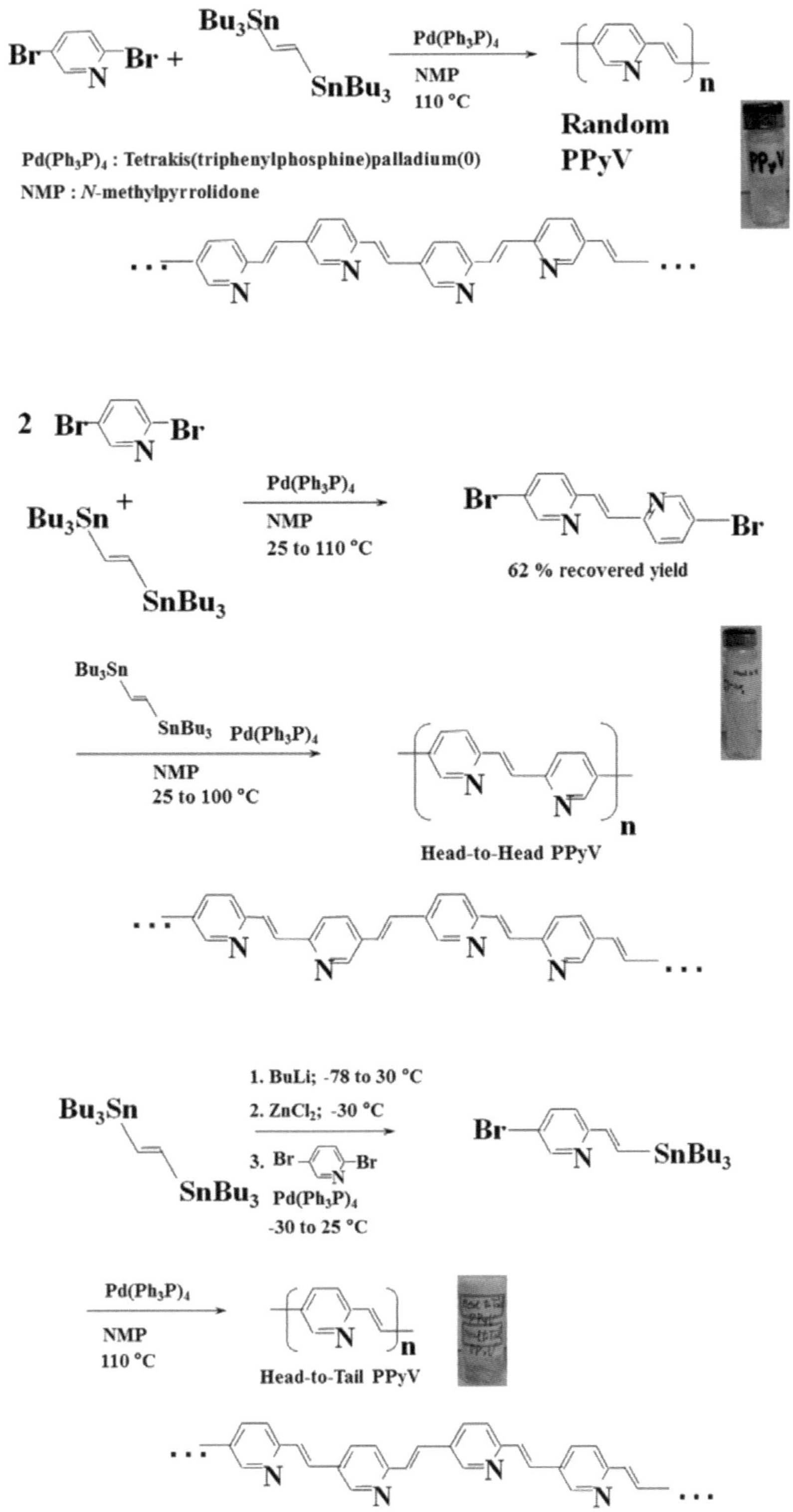

Figure 4. Schematic diagram of the synthetic route to Head-Head, Head-Tail, and Random PPyVs.

2.2. Characterization of PPyV and PMePyV

The PPyV appears a yellowish-orange powder and is not soluble in conventional organic solvents such as chloroform, methylene chloride, *etc*. However, the polymer is soluble in *m*-cresol, concentrated sulfuric acid and formic acid. These polymer solutions are highly fluorescent. The number-average molecular weight of PPyV was determined by gel permeation chromatography (GPC) against polystyrene standards to be about 7,300.

PMePyV is slightly soluble in water, but demonstrates high solubility in dimethyl sulfoxide (DMSO), N,N-dimethyl- formamide (DMF), and 1-methyl-2-pyrolidone (NMP). At room temperature, DMSO and NMP solutions of PMePyV rapidly turn dark, while after several hours DMF solutions gradually darken. The decomposition of the polymer may occur because of the nucleophilic attack of the solvent

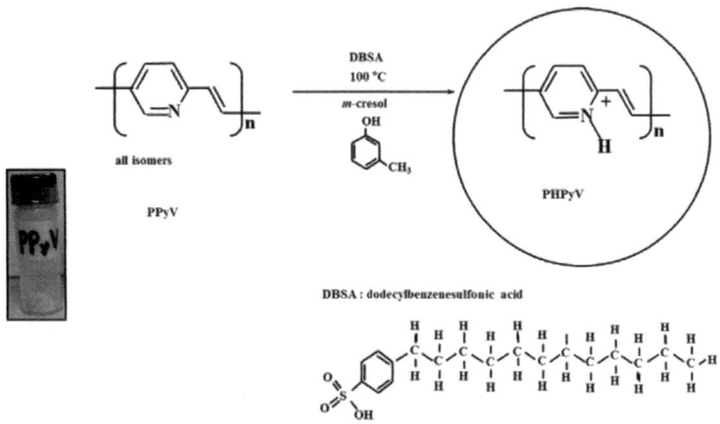

Figure 5. Conversion reaction of PPyV to yield PHPyV. The PPyV to the yellow-orange protonated poly(*p*-pyridyl vinylene), PHPyV is shown.

Protonated poly(*p*-pyridyl vinylene), PHPyV is positively charged polyelectrolyte like PMePyV and is prepared by treating PPyV with dodecylbenzenesulphonic acid at 100 °C for several days under argon (Ar), as shown in figure 5. Both pyridinium polymer, PMePyV and PHPyV are soluble in common organic solvents, such as chloroform and acetonitrile. PHPyV is yellow-orange in color.

On the other hand, about 20 years ago, a new technique of constructing multilayer assemblies by consecutively alternating adsorption polyelectrolyte was developed [14]. That is, the layer-by-layer molecular self-assembly technique of conductive polymers is based on the spontaneous adsorption of a charged polymer onto a surface from a dilute polymer solution. Both PHPyV and PMePyV are positively charged conjugated polyions. These polyelectrolytes can be fabricated into thin self-assembled films via alternating layers of positively and negatively charged polymers. This techniques makes it possible to readily prepare heterostructures with other conjugated or non-conjugated polyions [15].

3. Results and discussions

3.1. Kekulé and non-Kekulé structures of reduced PMePyV

PMePyV, when reduced, should be highly conductive as it is isoelectronic with the highly conductive n-doped poly(*p*-phenylene vinylene), PPV as shown in figure 6. However, what is not analogous is the fact that the reduced form of PMePyV bears no charge as shown in figure 7: all other known conductive polymers in their doped state bear charge [6]. Such an uncharged highly conductive polymer can be described as an intrinsic conductor. Carriers may be restricted by electrostatic

Journal of Physics: Conference Series **704** (2016) 012001 doi:10.1088/1742-6596/704/1/012001

interactions with dopant ions existing in an imperfect lattice. Accordingly as shown in figure 8, the content of three regioisomeric forms to have a significant effect on the polymer properties has to be considered, since reduction produces a closed shell state for the Head-Head linkages called Kekulé structure and an open shell state for Head-Tail linkages called non-Kekulé structure [8], [16].

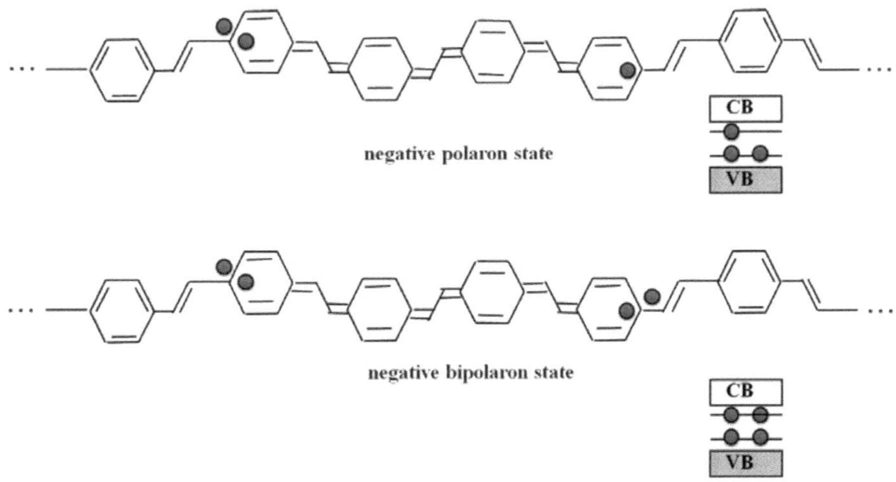

Figure 6. Electronic state of n-doped poly(*p*-phenylene vinylene), PPV.

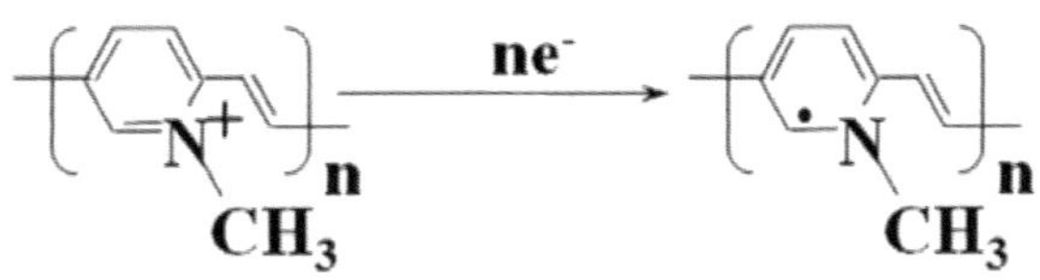

Figure 7. Reduced form of PMePyV.

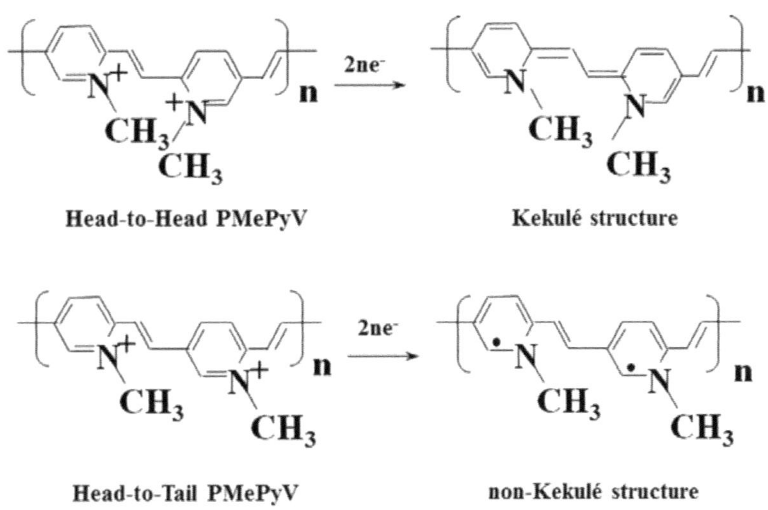

Figure 8. Schematic illustration of Kekulé and non-Kekulé states of reduced PMePyV.

3.2. Optical properties of PPyV, PMePyV and PHPyV

Figure 9 shows the optical absorption spectra of PPyV, PMePyV, and PHPyV. In PPyV, only a single absorption peak (λ_{ABmax}) was observed at about 420 nm due to interband excitation, and the band-gap energy estimated from the absorption edge is about 2.4 eV. Similar optical behaviors are also presented by PMePyV and PHPyV. λ_{ABmax}s of these polymers corresponding to the interband transition were observed at around 390 nm and 460 nm, respectively and the band-gap energy about 2.8 eV and 2.4 eV, respectively. This band-gap energy in PMePyV, 2.8 eV may be interpreted in terms of change of effective conjugation length due to the introduction of torsion of bonds between nitrogen rings on account of the steric hindrance between N atom and methyl chain.

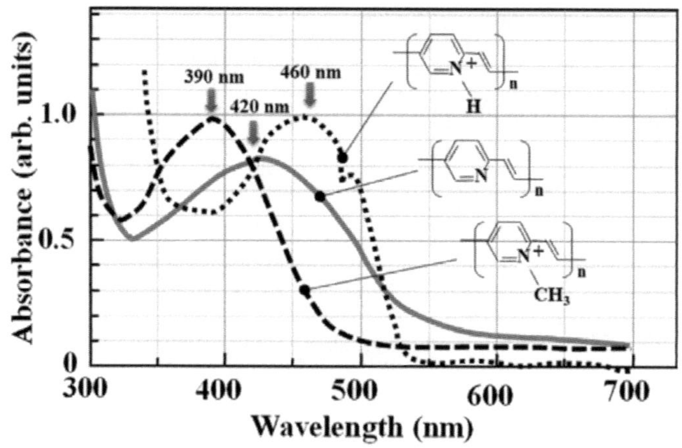

Figure 9. Optical-absorption spectra of PPyV, PMePyV and PHPyV.

On the other hand, the photoluminescence (PL) spectra of PPyV, PMePyV and PHPyV are shown in figure 10. For PL spectra these polymers were excited with light at 420 nm. The maximum PL peak wavelength (λ_{PLmax}) of PPyV and PMePyV shows peaks of 610 nm and 540 nm, respectively.

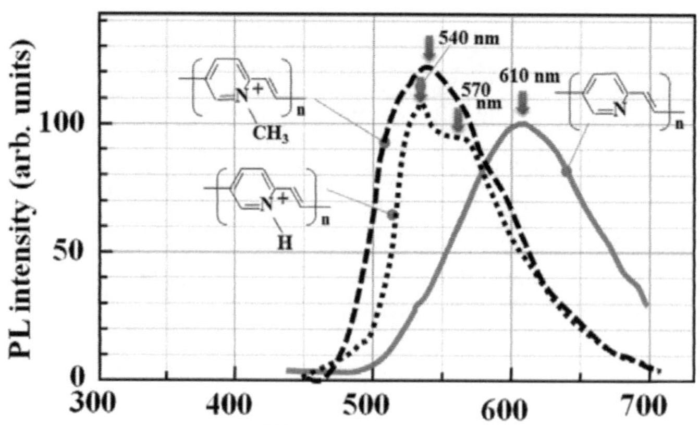

Figure 10. Photoluminescence spectra of PPyV, PMePyV and PHPyV.

As evident from figures 9 and 10, both the optical absorption (λ_{ABmax} at 390 nm) and PL (λ_{PLmax} at 540 nm) spectra of PMePyV are blue-shifted relative to PPyV to lower the energy of the filled

IOP Publishing
doi:10.1088/1742-6596/704/1/012001

orbitals significantly, this blue-shift may be affected by the methyl substitution too. On the other hand, the situation for PHPyV is different from that observed for PMePyV. Interestingly, the optical-absorption spectrum of PHPyV is red-shifted relative to PPyV, while the PL spectrum has two peaks evolved at around 540 nm and 570 nm. Both these two features are blue-shifted relative to PPyV. Though the mechanism of this anomaly is not clear at this stage, however the degree and uniformity of the protonation reaction may influence the PL spectrum. Detailed experimental studies of these unique phenomena will be carried out in the future.

Moreover, unfortunately, the significant difference was not recognized in optical properties for three regioisomers of PPyV, PMePyV and PHPyV. Though the detailed mechanism of this unique phenomenon is not clear at this stage, probably it is considered that optical measurements such as absorption and photoluminescence spectra are relatively structure insensitive properties.

3.3. Electrical conductivities of three regioisomeric PMePV

Electrical conductivity measurements have not been carried out *in situ* during electrolysis doping, because of loss of any reversible electrolysis behavior of PMePyV. In other word, reduced state of three regioisomeric forms was not so stable and the evaluation of intrinsic conductivity of these polymers was a little bit difficult, especially the temperature dependence of conductivity. Therefore, maximum electrical conductivities measured at room temperature for the reduced forms with three regioisomers of PPyV and PMePyV are summarized in Table 1.

Table 1. Electrical conductivities for the reduced forms with three regioisomers of PPyV and PMePyV.

	Random (S/cm)	**Head-to-Head (S/cm)**	**Head-to-Tail (S/cm)**
PPyV	2.8×10^{-3}	4.2×10^{-4}	1.8×10^{-2}
PMePyV	1×10^{-7}	1×10^{-7}	2.4×10^{-5}

at room temperature

The PPyV and PMePyV polymers are easily reduced and resisted oxidation. The electrochemical cycle between the desired doping potential and the neutral state was repeated for 2 or 3 times to obtain more uniformly doped samples. The doping conditions such as the film thickness (approximately 200 nm) and the composition of electrolyte (0.1 mol/L Bu_4NPF_6 in THF), are almost the same as electrochemistry explained in 3.4. The films thus obtained were washed immediately with THF for several times and cut into appropriate pieces to measure the electrical conductivity. These procedures were carried out in an Ar–filled dry box without exposure to air. The electrical conductivity was measured using either a standard four-probe or two-probe technique in an evacuated chamber.

The conductivity of Head-Tail PPyV shows one to two order of magnitude higher than those of both Random and Head-Head PPyV. On the other hand, methylation to form PMePyV caused a remarkable decrease of the conductivity. However, the conductivity of Head-Tail PMePyV shows a high value compared with those of other two regioisomeric forms. These results reflect both "carrierless" reduced forms of Head-Head PMePyV and the destruction of the highly reactive carriers generated on reduction of the Head-Tail PMePyV. A similar consideration may be observed for the conductivity of Random PMePyV. Moreover, the high conductivity of Head-Tail PMePyV and Head-Tail PPyV may be related to their non-Kekulé structure as shown in figure 8, if those are stable at room temperature. However, lengthy and careful consideration should be given to the origin of these unique phenomena.

3.4. Electrochemistry of random PMePyV and its regioisomers

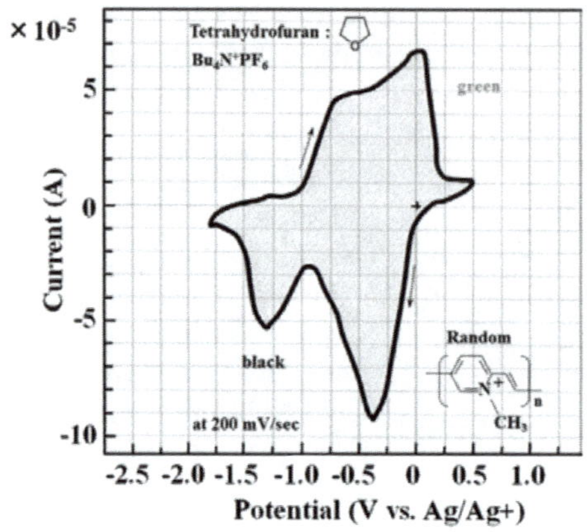

(a) Random PMePyV

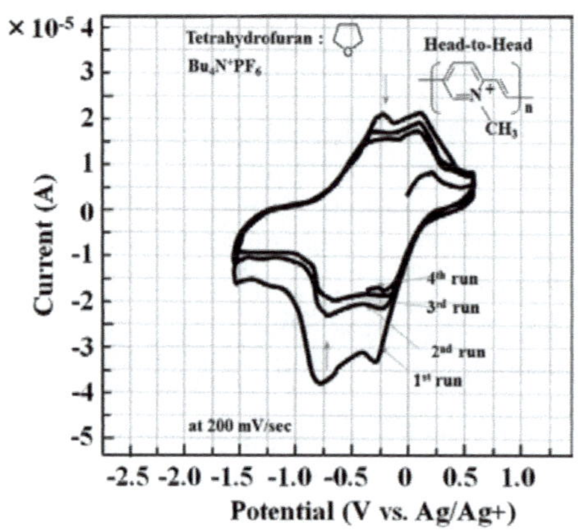

(b) Head-to-Head PMePyV

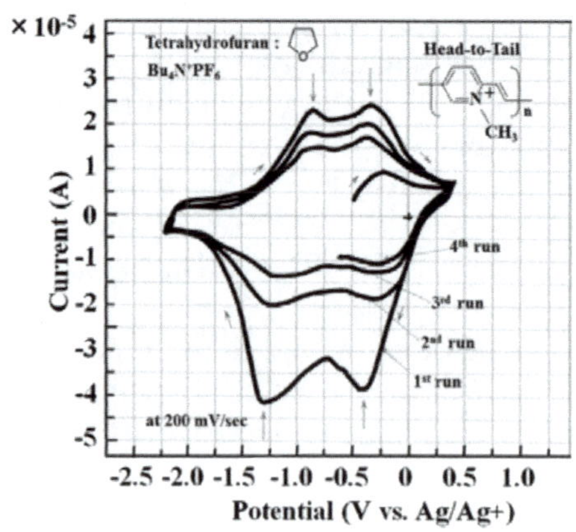

(c)Head-to-Tail PMePyV

Figure 11. Cyclic voltammetry of the three regioisomeric forms of (a) Random, (b) Head-to-Head, and (c) Head-to-Tail PMePyV.

The cyclic voltammetry of three regioisomeric forms of PMePyV during electrochemical redox reaction was studied using a three-electrode cell in which the working electrode consisted of an about 200 nm thick film of samples deposited on an ITO (indium-tin oxide) glass substrate. The counter and reference electrodes were Pt plate and Ag wire, respectively. The composition of electrolyte used for the cyclic voltammetry is 0.1 mol/L tetrabutylammonium hexafluorophosphate (Bu_4NPF_6) as the supporting electrolyte in THF. In this study, the reduction of these polymers were easy, but the oxidation difficult.

As shown in figure 11 (a), the cyclic voltammogram of the Random PMePyV demonstrated high coulombic reversibility. Two well defined reduction waves are centered at about -0.35 and -1.26 volts vs. Ag/Ag^+. And then, two oxidation waves slightly overlap, showing current peaks at -0.6 and 0 volts.

On the other hand, cyclic voltammetry of both the regiochemically Head-Head and Head-Tail PMePyV show a ra pid decrease in activity after the first reduction cycle. As shown in figure 11 (b), Head-Head PMePyV shows two reduction waves centered at about -0.25 and -0.75 volts. And two oxidation waves which show considerably less coulombic activity than their corresponding reduction waves are centered at about -0.25 and +0.12 volts. A similar cyclic behavior is also presented by Head-Tail PMePyV. As shown in figure 11 (c), two reduction waves centered at about -0.4 and -1.3 volts. And two oxidation waves which show considerably less coulombic activity than their corresponding reduction waves are centered at about -0.3 and +0.8 volts. The continued cycling of these two PMePyV resulted in loss of electrolysis activity.

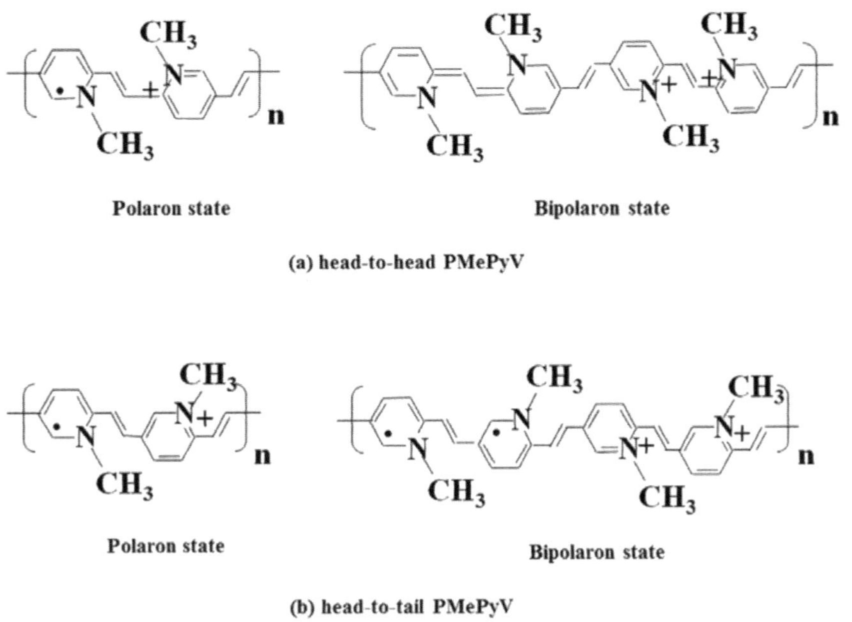

Figure 12. Speculated polaron and bipolaron states of (a) Head-Head and (b) Head-to-Tail PMePyV generated on reduction.

In Head-to-Head and Head-to-Tail PMePyV, though the cause of the inactivating phenomenon by the repetition of the cyclic voltammetry measurement is not clear at this stage, the Kekulé and non-Kekulé structures of reduced regioisomeric form described in figure 8 might also contribute to the loss of reversible electrochemical behavior. Because, for example, polaron and bipolaron states of (a) Head-to-Head and (b) Head-to-Tail PMePyV generated on reduction is estimated as shown in figure

12, the non-Kekulé structures are formed and the high reactivity of them may affect to the physical properties of regioisomeric forms of PPyV.

4. Conclusion

In order to clarify the stability of n-type conductive polymers, fundamental physical properties of three regioisomeric forms of PPyV (Random, Head-Head, and Head-Tail) have been investigated. The structural change (reduced states) of Head-Head PMePyV and Head-Tail PMePyV, that is, Kekulé and non-Kekulé structures, may strongly affect to the stability.

Reference

[1] Dennard R H, Gaensslen F H, Rideout V L, Bassous E and LeBlanc A R 1974 *IEEE Journal of Solid-State Circuits* **9**(5) 256-268
[2] Galarini R, Musco A, Pontellini R, Bolognesi A, Destri S, Catellani M, Mascherpa M and Zhuo M 1991 *Journal of the Chemical Society, Chemical Communications* (6) 364-365
Aviram A and Ratner M A 1974 Chemical Physics Letters **29**(2) 277-283
[3] Onoda M, Kaneto K, Ohsawa T and Yoshino K 2016 *Organic Iontronics* (Tokyo: Morikita Publishing Co., Ltd.) in Japanese
[4] Onoda M (Supervisor) 2012 *Physical properties and evaluation of conductive polymers for high performance organic electronic devices* (Tokyo-Osaka: CMC Publishing Co., Ltd.) in Japanese
[5] Onoda M 2013 in *New Developments in Polymer Composites Research* (edited Laske S and Witschnigg A New York: Nova Publishers) 67-88
[6] Yoshino K and Onoda M 1996 *Polymer Electronics* (Tokyo: Corona Publishing Co., Lted.) in Japanese
[7] de Leeuw D M, Simenon M M J, Brown A R and Einerhand R E F 1997 *Synthetic Metals* **87** (1) 53-59
[8] Marsella M J, Fu D-K and Swager T M 1995 *Advanced Materials* **7**(2) 145-147
[9] Galarini R, Musco A, Pontellini R, Bolognesi A, Destri S, Catellani M, Mascherpa M and Zhuo M 1991 *Journal of Chemical Society, Chemical Communications* (6) 364-365
[10] Onoda M 2015 *International Conference & Exhibition on Advanced & Nano Materials* OPHM-1
[11] McCullough R D, Lowe R D, Jayaraman M and Anderson D L 1993 *The Journal of Organic Chemistry* **58**(4) 904-912
[12] Stille J K 1986 *Angewandte Chemie International Edition in English* **25**(6) 508-524
[13] Beddoes R L, Cheeseright T, Wang J and Quayle P 1995 *Tetrahedron Letters* **36**(2)283-286
[14] Decher G, Hong J D and Schmitt J 1992 *Thin Solid Films* **210-211**(2) 831-835
[15] Onoda M 2013 in *New Developments in Polymer Composites Research* (edited Laske S and Witschnigg A New York: Nova Publishers) 89-149
[16] Vollhardt K P C and Schore N E 1999 *Organic Chemistry, Structure and Function* 3rd Edition (New York and Oxford: W. H. Freeman and Company)

Journal of Physics: Conference Series **704** (2016) 012002 doi:10.1088/1742-6596/704/1/012002

Parametric Optimization of Experimental Conditions for Dye-Sensitized Solar Cells based on Far-red Sensitive Squaraine Dye

Takuya Morimoto, Naotaka Fujikawa, Yuhei Ogomi, Shyam S. Pandey, Shuzi Hayase

Graduate School of Life Science and Systems Engineering, Kyushu Institute of Technology, 2-4 Hibikino, Wakamatsu, Kitakyushu 808-0196, Japan

E-mail: morimoto-takuya@edu.life.kyutech.ac.jp; shyam@life.kyutech.ac.jp

Abstract. A far-red sensitive unsymmetrical squaraine dye SQ-41 has been synthesized and subjected to the fabrication of dye-sensitized solar cells by varying the various parameters in order attain optimum photoconversion efficiency (η). It has been demonstrated that an optimum ratio of dye to coadsorber, thickness of mesoporous TiO_2 layer, redox electrolyte and surface treatment are necessary to enhance overall external η. In the case of surface treatment, it has been shown to exhibit pronounced device performance when both of the FTO as well mesoporous TiO_2 surfaces were treated with aqueous $TiCl_4$. In spite of very high molar extinction coefficient of dye SQ-41, 10-12 μm thickness of mesoporous TiO_2 was found to be necessary to attain the maximum η.

1. Introduction

Dye sensitized solar cells (DSSCs) consisted of dye-adsorbed mesoporous TiO_2 as photoanode, Pt coated transparent conductive oxide (TCO) glass as cathode and a suitable electrolyte have been emerged as low cost next generation solar cells with minimal adverse impact on environment during its fabrication [1]. Research and development on various aspects of DSSCs in the last about 20 years led to the achievement of photoconversion efficiency (η) over 11 % even with sensitizers having photon harvesting mainly in the visible region of the solar spectrum [2-3]. Solar spectrum is consisted of photons from varying wavelengths encompassing from ultraviolet to infra-red (IR). Nearly quantitative photon harvesting in the visible wavelength reason has already been achieved by potential organic sensitizers indicating the need for development of suitable sensitizing dyes having efficient photon harvesting in the higher wavelength region [4-5]. To enhance the external power conversion efficiency even higher, it is highly desired to develop novel and efficient sensitizing dyes that can harvest photons from far-red to NIR wavelength region. Utilization of such dyes in combination with already developed as well as commercially available potential visible light absorbing sensitizers in hybrid or tandem DSSC architecture is expected to enhance the overall efficiency of the DSSCs.

In this context, squaraine dyes serve as one of the model representatives fulfilling this demand owing to very sharp and narrow light absorption that can be tuned from visible to IR wavelength region by judicious molecular design [6]. In past two decades after inception of DSSCs, a lot of works have been conducted for optimizing various DSSC fabrication parameters using visible light absorbing dyes in order to enhance the η [7]. NIR dyes are typically different as compared to their visible light absorbing dye counterparts due to narrow full width at half maximum and very high molar extinction coefficients. Therefore, such optimization parameters can be partially applied for NIR dyes due to their intense light absorption and enhanced aggregation behaviour owing to their relatively planer molecular structure and extended π-conjugation. This article deals with the influence of various DSSC fabrication parameters such as thickness of mesoporous TiO_2 layer, surface treatment, concentration of the dye-aggregating agent chenodeoxycholic acid (CDCA) and nature of redox electrolyte upon the photovoltaic behaviour using a model far-red squaraine dye SQ-41. Structures of dye SQ-41 and CDCA have been shown in the inset of Fig. 1.

2. Materials and Methods

Far-red sensitive unsymmetrical squaraine dye (SQ-41) has been synthesized and its structure has been characterized as per our earlier publication [8]. Electronic absorption spectra of the dye in ethanol solution as well as thin film of dye absorbed on mesoporous TiO_2 has been measured using UV-visible spectrophotometer (JASCO, V670). DSSC was fabricated using Ti-nanoxide D/SP and HT/SP pastes (Solaronix) on fluorine doped tin-oxide (FTO) glass by doctor blading followed by backing at 450°C for 30 min. This coating and backing process was repeated multiple times to prepare TiO_2 coated FTO substrates of variable thicknesses (2 - 20 μm). In order to optimize the surface properties of FTO and TiO_2 layers, $TiCl_4$ surface treatment was also conducted. This surface treatment was performed by dipping the substrates in 40 mM solution $TiCl_4$ in water at 70°C for 30 minutes followed by the backing at 500°C for 30 min. This makes a very thin and compact TiO_2 layer on the substrates. The TiO_2 coated FTO substrate thus obtained was subjected to dye adsorption at room temperature for 4 hours. Dye bath solution for dye adsorption was prepared by making 0.2 mM dye solution in ethanol using 10 mM of CDCA as dye aggregation preventing agent. This optimum ratio of dye to CDCA was selected based on our previous publication using this dye [9]. A thin catalytic layer of Pt was coated on FTO glass to use as counter electrode. Three different kind of iodine based redox electrolytes (**A, B** and **C**) have been used to see the implication of the constituents and their relative concentration as per the recipe in the reported literatures [10-12].

Electrolyte **A** was consisted of I_2 (0.05 M), LiI (0.5 M), 1,2-dimethyl-3-propylimidazolium iodide (0.6M) and t-butylpyridine (TBP, 0.5 M) in acetonitrile [10]. Electrolyte **B** was consisted of I_2 (0.05 M), LiI (0.5 M), 1,2-ethyl-3-methylimidazolium dicyanamide (0.6M) and TBP (0.6 M) in acetonitrile [11]. Electrolyte **C** was consisted of I_2 (0.05 M), LiI (0.5 M), 1,3-dimethylimidazolium iodide (0.5M), Guanadinium thiocyanate (0.1 M) and TBP (0.5 M) in a mixture of acetonitrile and valeronitrile (85:15) [12]. A 25 μm thick Himilan film (Mitsu-Dupont) was used as hot melt spacer. Device area of (0.2025 cm^2) was precisely controlled using a black metal mask on the DSSC during the photovoltaic measurements every time. Photovoltaic performance of DSSC thus fabricated was measured with a solar simulator (CEP-2000, Bunko Keiki, Japan) equipped with Xenon lamp for the light exposure. The spectrum of the solar simulator and its power were adjusted to be at global air mass 1.5 condition (100 mW/cm^2) using a spectroradiometer (LS-100, Eiko Seiki, Japan). Photocurrent action spectra of the solar cells were also measured with a constant photon flux of 1 x 10^{16} photons/cm^2 using an action spectrum measurement system connected to the solar simulator.

3. Results and Discussion

3.1. Photophysical Characterization

Figure 1 shows the electronic absorption and fluorescence emission spectra of sensitizing dye SQ-41 in ethanol solution along with the thin film adsorbed on mesoporous TiO_2. Electronic absorption spectrum shows a sharp absorption maximum (λ_{max}) at 634 nm with very high molar extinction coefficient of 1.98 x 10^5 $dm^3.mol^{-1}.cm^{-1}$ which is associated with the π-π* electronic transition with a small vibronic shoulder around 590 nm. At the same time, emission spectrum shows slightly red-shifted emission maximum at 642 nm having very small Stoke shift of 8 nm. This spectral behaviour is a typical characteristic of squaraine dyes and very small Stoke shift indicates the conformational rigidity of this molecule in the excited state. On the other hand, electronic absorption spectrum of this dye in solid state shows a little bathochromically shifted λ_{max} of 647 nm along with spectral broadening. This is attributed to the interaction of dye with TiO_2 in the condensed state.

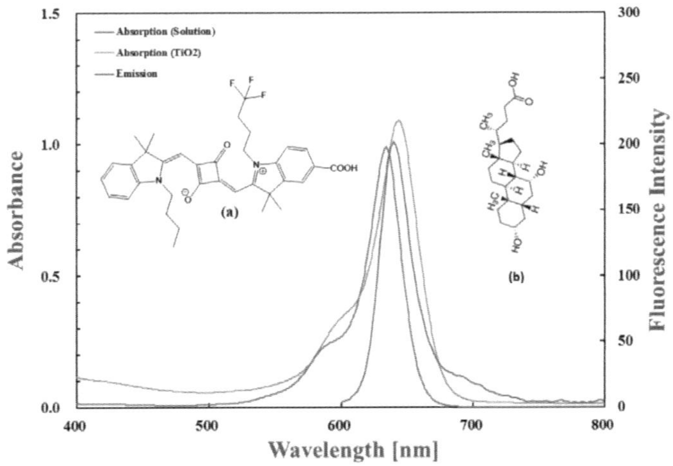

Figure 1. Electronic absorption spectra of SQ-41 in ethanol solution (5 μM), thin film on mesoporous TiO_2 and fluorescence emission spectrum in ethanol solution (1 μM). Inset shows the molecular structure of sensitizer SQ-41(a) and coadsorber CDCA (b).

3.2. Influence of $TiCl_4$ Surface Treatment

Photoanode in DSSC is consisted of FTO/TiO_2/dye interfaces and is the most important component of DSSCs which must be optimised for optimum device performance. Actually mesoporous TiO_2 is composed of interconnected nanoparticles which provides very high internal surface area promoting the high dye loading. At the same time, these interconnected TiO_2 nanoparticles offer not only limitation to facile electron transport but also charge recombination with electrolyte ions also. Since it is mesoporous in nature and electrolyte ions are also in contact with FTO surface, it offers the possibility of charge recombination between FTO and electrolyte ions also. To circumvent these problems application of a very thin layer of compact TiO_2 on FTO as well as TiO_2 has been found to be beneficial and used by many DSSC researchers. Surface treatment using aqueous $TiCl_4$ followed by backing is one of the most commonly used methods to prepare very thin compact layer of TiO_2. Apart from charge recombination blocking layer, this compact TiO_2 layer has been reported to play various roles like to increase the TiO_2 surface area [13], to enhance electron transport [14], to promote light scattering [15] and dye anchoring [16] finally leading to enhanced overall device performance. In this work, attempts have been made to investigate the role of this compact layer on overall device performance and demonstrate that which one plays the predominant role. In order to predict this effect more explicitly, other factors such as thickness of mesoporous TiO_2 layer (12-14 μm), dye (SQ-41), dye/CDCA ratio (1/50) and electrolyte (Electrolyte B) were kept the same.

Figure 2 shows the current-voltage (I-V) characteristics of DSSCs based SQ-41 having different surface treatment conditions under simulated solar irradiation along with the photovoltaic parameters in terms of short circuit current density (J_{sc}), open circuit voltage (V_{oc}), fill factor (FF) and η are summarized in the table 1. It can be clearly seen from this figure and table 1 that surface treatment of FTO has only little effect on η while surface treatment on mesoporous TiO_2 layer leads to drastic improvement of J_{sc} from 4.15 mA/cm^2 to 7.04 mA/cm^2 as compared to that obtained for without any surface treatment. Best DSSC performance with external power conversion efficiency 3.54 % was obtained when both of the FTO and TiO_2 surfaces were subjected to $TiCl_4$ surface treatment. In order to explain this profound enhancement of Jsc upon surface treatment, photocurrent action spectra also known as plot of incident photon to current conversion efficiency (IPCE) as a function of wavelength of monochromatic light irradiation was also measured and has been shown in the Fig. 2. Action spectra shown in Fig. 2 clearly reveal the far-red photon harvesting by sensitizers and maxima of the observed IPCE is in accordance with the Jsc measured in the I-V characteristics.

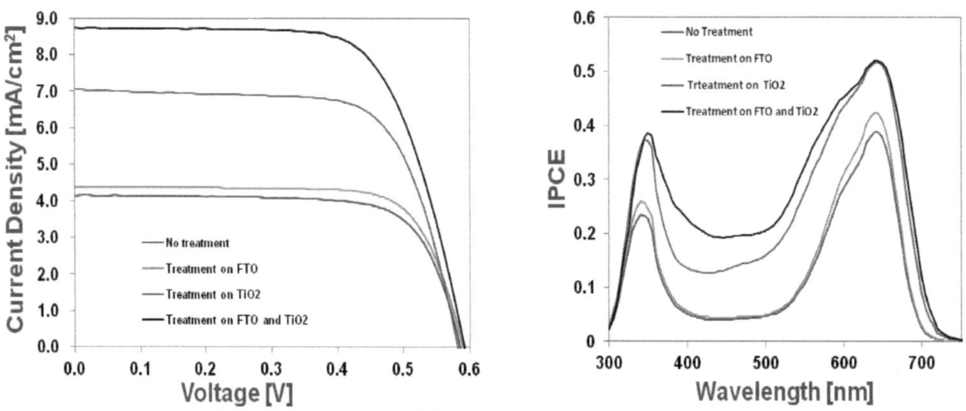

Figure 2. Photovoltaic characteristics of DSSCs under simulated solar irradiation (left) and photocurrent action spectra (right) after monochromatic incident light for DSSCs with different types of $TiCl_4$ surface treatments.

Table 1. Photovoltaic parameters for DSSC based on SQ-41 having $TiCl_4$ surface treatments conducted on different surfaces of photoanode.

Treatment Conditions	Efficiency (%)	FF	V_{oc} (V)	J_{sc} (mA/cm^2)
No surface Treatment	1.79	0.74	0.58	4.15
Treatment on FTO glass only	1.93	0.75	0.59	4.37
Treatment on TiO_2 layer only	2.88	0.70	0.58	7.04
Treatment on FTO glass and TiO_2 layer both	3.54	0.69	0.59	8.74

3.3. Thickness of Mesoporous TiO_2 Layer

Sensitizing dyes play most important role of photon harvesting in DSSCs. In order to have sufficient photon harvesting, mesoporous TiO_2 has been most extensively used for adsorption of dyes owing to its wide band gap and very high surface area. To achieve the best performance of DSSC an optimum thickness of the TiO_2 layer is highly desired. Lower thickness leads to non-optimal dye loading while too high thickness hampers the diffusion of redox species through the nanopores promoting the chances of charge carrier recombination. This optimal thickness depends on the nature of dye as well as particle size and preparation conditions of the mesoporous TiO_2. Using Ruthenium complex based dyes (N-719) Kang et al have demonstrated the need of 15 μm thick TiO_2 layer for best efficiency [17]. On the other hand, Yamaguchi et al used 34 μm thick mesoporous TiO_2 for the optimum efficiency of DSSC using Black dye [18]. Thus an optimization of TiO_2 layer thickness especially for this class of dyes is highly desired owing to their very high molar absorption coefficient where relatively thinner films are capable of absorbing sufficient photons. Keeping this in mind, DSSCs were fabricated by varying the thickness of mesoporous TiO_2 layer keeping other variables like dye, dye concentration, surface treatment, electrolyte (Electrolyte B), CDCA concentration etc. constant. Fig. 3 exhibits a typical I-V characteristic of the DSSC having the TiO_2 layer thickness of 10 μm giving a power conversion efficiency of 3.39 % under simulated solar irradiation. Photovoltaic characteristics of the DSSCs for other thicknesses of

TiO_2 were also measured and photovoltaic parameters in terms of J_{sc}, V_{oc}, FF and η are summarized in the table 2. A perusal of the inset of Fig. 2 indicating thickness dependence of η reveals that increase in the thickness leads to fast increase in the η up to about 10-14 μm and starts a sharp decline after 14 μm. This indicates at least 10 μm of mesoporous TiO_2 thickness is needed to achieve optimum efficiency in spite of such a high molar extinction coefficient of this dye.

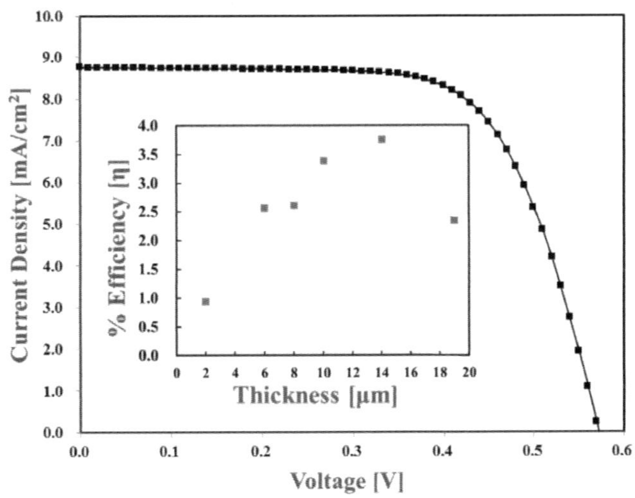

Figure 3. Photovoltaic characteristic of DSSC with TiO_2 layer thickness of 10 μm under simulated solar irradiation. Inset shows the thickness dependence of photoconversion efficiency

Table 2. Photovoltaic parameters for DSSC based on SQ-41 with varying thickness of the mesoporous TiO_2 layer.

Thickness	2 μm	6 μm	8 μm	10 μm	14 μm	19 μm
Efficiency (%)	0.94	2.57	2.61	3.39	3.76	2.35
FF	0.72	0.61	0.62	0.68	0.67	0.62
V_{oc} (V)	0.55	0.57	0.56	0.57	0.59	0.58
J_{sc} (mA/cm^2)	2.40	7.35	7.51	8.76	9.58	6.52

3.4. Influence of Nature of Electrolyte

Redox electrolyte plays a vital role in the dye regeneration during DSSC operation and affects all of photovoltaic parameters like J_{sc}, V_{oc} and FF finally controlling the η. Amongst several discovered electrolytes for DSSCs till date, iodine based redox couple I^-/I_3^- are still preferred for research laboratory and exhibited its dominance until 2010 [19]. This electrolyte works well with approximately all kinds of sensitizing dyes used in DSSCs. It is unlike to cobalt complex based redox shuttles needing specific dye design especially dyes bearing larger number of long alkyl chains [20]. It mainly contains the sources of I^- and I_3^- to form redox couple, I^- supplying ionic liquid for long term stability, some additives to tune the energetics of mesoporous TiO_2 and finally a suitable solvent. In this investigation, three different recipes of iodine based redox electrolyte (Electrolyte **A**, **B** and **C**) which have been most extensively utilized in DSSC research for Ru dyes have been attempted and their influence on photovoltaic performance was explored. In order to predict the main role played by electrolytes and its constituents, other DSSC fabrication parameters like thickness of mesoporous TiO_2 layer (12-13 μm), dye (SQ-41), CDCA to dye ratio (50), $TiCl_4$ surface treatments etc. were fixed varying only the type of electrolyte. Figure 4 depicts the I-V characteristics of DSSCs using different types of electrolyte bearing I^-/I_3^- redox couple along with photovoltaic parameters shown in the table 3. A perusal of Fig. 4 and table 3 clearly corroborates that even under similar other DSSC fabrication parameters, only change in the recipe and components of iodine based redox electrolyte affects the overall photovoltaic performance. Only change in the concentration of TBP as additive and nature of ionic liquid for electrolyte **A** leads to enhancement of both of the observed J_{sc} and V_{oc} as compared to electrolyte **B** suggesting the importance of judicious selection of particular electrolyte system in order to achieve the optimum photo conversion efficiency.

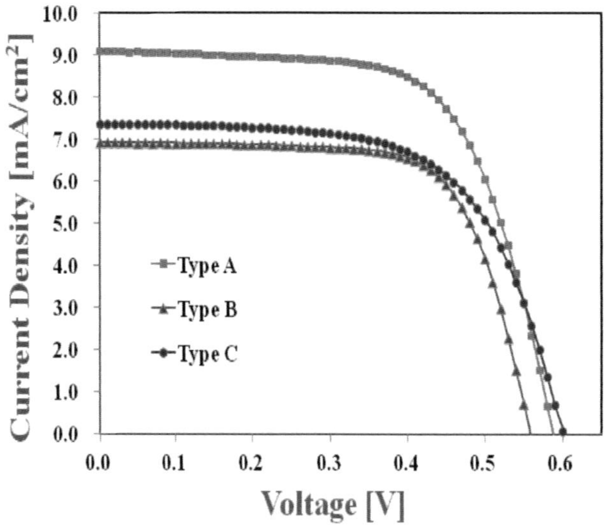

Figure 4. Photovoltaic characteristics of DSSC using SQ-41 in the presence of different iodine based redox electrolyte.

Table 2. Photovoltaic parameters for DSSC based on SQ-41 with different types of I^-/I_3^- based redox electrolytes

Electrolytes	Type-A	Type-B	Type-C
Efficiency (%)	3.50	2.70	2.77
FF	0.65	0.70	0.63
V_{oc} (V)	0.59	0.56	0.60
J_{sc} (mA/cm^2)	9.10	6.94	7.37

4. Conclusions

It has been demonstrated that control and optimization of DSSC fabrication parameters are highly desired in order to achieve optimum photoconversion efficiency for a particular class of dye sensitizer. Utilization of CDCA as coadsorber leads to enhanced photoconversion efficiency and 50-100 times of CDCA with respect to dye concentration was found to be optimum. $TiCl_4$ surface treatment was found to control the overall photoconversion efficiency and as compared to FTO layer, $TiCl_4$ treatment of mesoporous TiO_2 layer was found to play dominant role to achieve the best efficiency. Finally mesoporous TiO_2 layer having thickness in the range of 10-14 μm and iodine based redox electrolyte **A** was found to be optimum for DSSCs using far-red sensitive SQ-41 dye as photosensitizer.

Acknowledgments
One of the authors (SSP) would like express sincere thanks to Japan Society for the Promotion of Science (JSPS) for the financial support by Grant-in-Aid for Scientific Research (C) (Grant Number 26410206).

References

[1] Grätzel M 2009, *Acc. Chem. Res.* **42** 1788
[2] Kakiage K, Aoyama Y, Yano T, Otsuka T, Kyomen T, Unno M and Hanaya M 2014 *Chem. Commun. 50 6379*
[3] Wang M, Grätzel C, Zakeeruddin S M, Grätzel M 2012 *Energy Environ. Sci.* **5** 9394
[4] Mathew S, Yella A, Gao P, Humphry-Baker R, Curchod BF, Ashari-Astani N, Tavernelli I, Rothlisberger U, Nazeeruddin M K, Gratzel M 2014 *Nat. Chem.* **6** 242
[5] Gao F, Wang Y, Shi D, Zhang J, Wang M, Jing X, Humphry-Baker R, Wang P, Zakeeruddin S M, Grätzel M 2008 *J. Am. Chem. Soc.* **130** 10720
[6] Sreejith S, Carol P, Chithra P, Ajayaghosh A 2008 *J. Mater. Chem.* **18** 264
[7] Ito S, Murakami T N, Comte P, Liska P, Grätzel C, Nazeeruddin M K, Grätzel M 2008 *Thin Solid Films* **516** 4613
[8] Pandey S S, Inoue T, Fujikawa N, Yamaguchi Y, Hayase S 2010 *Thin Solid Films.* **519** 1066
[9] Morimoto T, Fujikawa N, Ogomi Y, Pandey S S, Ma T, Hayase S 2016 *J. Nanosci. Nanotechnol.* (In press)
[10] Hara K, Tachibana Y, Ohga Y, Shinpo A, Suga S, Sayama K, Sugihara H, Arakawa H 2013 *Sol. Energy Mater. Sol. Cells* **77** 89
[11] Yoshida Y, Tokashiki S, Kubota K, Shiratuchi R, Yamaguchi Y, Kono M, Hayase S 2008 *Sol. Energy Mater. Sol. Cells* **92** 646
[12] Gao F, Wang Y, Shi D, Zhang J, Wang M, Jing X, Humphry-Baker R, Wang P, Zakeeruddin S M, Grätzel M 2008 *J. Am. Chem. Soc.* **130** 10720
[13] Barbe C J, Arendse F, Comte P, Jirousek M, Lenzamann F, Shklover V, Grätzel M 1997 *J. Am. Ceram. Soc.* **80** 3157
[14] Sommeling P M, O'Regan B C, Haswell R R, Smit H J P, Bakker N. J, Smits J J T, Kroon J M, van Roosmalen J A M 2006 *J. Phys. Chem.* B **110** 19191
[15] Park N G, Schlichthorl G, Lagemaat J V D, Cheong H M, Mascarenhas A, Frank A J 1999 *J. Phys. Chem. B* **103** 3308
[16] Zeng L Y, Dai S Y, Wang K J, Pan X, Shi C W, Guo L 2004 *Chin. Phys. Lett.* **21** 835
[17] Man Gu Kang, Kwang Sun Ryu, Soon Ho Chang, Nam Gyu Park, Jin Sup Hong, Kang-Jin Kim 2004 *Bull. Korean Chem. Soc.* **25** 742
[18] Yamaguchi T, Uchida Y, Agatsuma S, Arakawa H 2009 *Sol. Energy Mater. Sol. Cells* **93** 733
[19] Jihuai W, Zhang L, Jianming L, Miaoliang H, Yunfang H, Leqing F and Genggeng L 2015 *Chem. Rev.* **115** 2136
[20] Zong X, liang M, Fan C, Tang K, Li G, Sun Z, Xue S 2012 *J. Phys. Chem. C.* **116** 11241

Tandem Dye-Sensitized Solar Cells Based on TCO-less Back Contact Bottom Electrodes

Ajay K. Baranwal*, Naotaka Fujikawa, Terumi Nishimura, Yuhei Ogomi, Shyam S. Pandey, Tingli Ma and Shuzi Hayase*

Division of Green Electronics, Graduate School of Life Science and Systems Engineering, Kyushu Institute of Technology, 2-4 Hibikino, Wakamatsu, Kitakyushu 808-0196, Japan

E-mail: ajaybarn@gmail.com ; hayase@life.kyutech.ac.jp

Abstract. Mechanically stacked and series connected tandem dye sensitized solar cells (T-DSSCs) are fabricated in novel architectures. The architecture consist of TCO tandem DSSCs stacked with TCO-less back contact DSSCs as bottom electrodes (TCO-less tandem DSSCs). Resulting TCO-less tandem DSSCs architecture finds its usefulness in efficient photon harvesting due to improved light transmission and enhanced photons reaching to the bottom electrodes. The fabricated tandem performance was verified with visible light harvesting model dyes D131 and N719 as a proof of concept and provided the photoconversion efficiency (PCE) of 6.06 % under simulated condition. Introduction of panchromatic photon harvesting by utilizing near infrared light absorbing Si-phthalocyanine dye in combination with the modified tandem DSSC architecture led to enhancement in the PCE up to 7.19 %.

1. Introduction

Mesoscopic injection solar cells (DSSCs) have emerged an important economical photovoltaic technology with associated advantages of flexibility, colorful and semitransparency. The superior performance at low light intensity make them valuable for interior decorative photovoltaic purposes [1]. Constituents of DSSCs include transparent conductive oxide (TCO) layer, mesoporous TiO_2 layer, sensitizers, electrolytes and catalytic layer. Intense efforts are carried out to enrich photoelectric conversion efficiency (PCEs) and Hanaya et al achieved a mile stone of more than 14% by designing novel sensitizers in spite of the photon harvesting mainly in the visible reason of the solar spectrum [2]. A panchromatic photon harvesting from visible to near infra-red region is, therefore, highly desired to enhance the photoconversion efficiency further. Panchromatic light absorption using a single near infra-red (NIR) dyes have been although attempted, however, TiO_2 based DSSCs show energetic matching limitations in terms of the electron injection and dye regeneration ultimately led to poor performance [3]. In order to circumvent this problem of single dye for panchromatic photon harvesting, use of multiple dyes either as dye-cocktail or positioning of dyes in dye bilayer approach have shown some success for efficiency enhancement in the recent past. In spite of some success using these approaches, constraints like inter-dye unfavorable interactions and complexity involved in selective positioning of dyes in bilayer architecture are still existing hurdles [4-5].

Another possible approach towards utilization of multiple dyes for panchromatic photon harvesting is the implementation of tandem device architecture. According to SQ limit the tandem approach evade the performance limitation established by single junction solar cells. The electrical and optical stacking of multiple DSSCs (tandem DSSCs) avoid the efficiency limitation imposed and can be realized either in series or parallel interconnection. Series arrangement needed the respective top cell and bottom cell (sub cell) short circuit current density (Jsc) matching emanating in the enhanced open circuit voltage (Voc). Mechanically stacked tandem DSSCs architecture possess numerous advantages due to ease of fabrication, independent control and optimization and top as well as bottom cells along with flexibility for selection and utilization of redox electrolytes. The first tandem DSSCs was explored and reported by He et al by utilizing n-p photoanodes [6]. Although realization of tandem DSSC was demonstrated by enhanced Voc but overlapped spectra of corresponding sub cells resulted in to < 1% external power conversion efficiency (PCE). Owing its simplicity and huge possibility of combining different types of top and bottom cells such as DSSC/DSSC [7, 8], DSSC/Copper-indium-gallium selenide (CIGS) [9-10], DSSC/Silicon [11], DSSC/Perovskite [12] and DSSC/bulk heterojunction [13] etc., series connected mechanically stacked tandem solar cells have been most extensively investigated aiming towards enhancement in the overall PCE by enhancing the Voc. In all these attempts, in spite of some success on the enhanced PCE due to Voc enhancement, bottle-neck still lies in the control and optimization of optical losses by intermediate layers, glass, and top cell hindering the photon reaching to the bottom cell. Especially in the case of DSSC/DSSC tandem cells, superfluous intermediate TCO layer used in the stacked tandem DSSCs are liable for loss of the photon flux accessible to bottom cell. It is reported that about 20% photons are lost while transmitting from the top cell and intermediate layers [14-15].

Intermediate layers have been found to play a crucial role towards the realization efficient tandem solar cell and consequence of their presence have shown to affect the photovoltaic performance adversely not only in the perovskite-Si but also perovskite-CIGS tandem solar cells [16-17]. To deal with such unavoidable losses by intermediate layers, efforts have been directed to fabricate n-p tandem solar cells utilizing p-NiO as photocathode. Conceptually, the n-p tandem architecture seems to be better than conventional n-n tandem design, since it not only omits intermediate glass layers but also there is no need of expensive Pt counter electrode and simplicity of the tandem design. Contrary to n-DSSC, p-DSSC has not got much attentions owing to their very low reported efficiency (<0.2 %) [18]. Even best p-DSSC developed so for, its implementation in n-p tandem DSSC has led to external PCE of < 2 % [19]. Apart from this other approaches in terms of alteration and modification of intermediate layers by utilizing Pt-coated metal mesh have also been attempted in parallel tandem DSSC configuration. 20-21]. Enhanced Jsc has been although demonstrated but such architecture are only valid to DSSC/DSSC tandem solar cells limiting its wide spread hybridization with other kind solar cells. Till date best PCE has been demonstrated by mechanically stacked series connected tandem solar cells but in most commonly employed tandem architecture four TCO plates are being used. It not only offer additional cost burden since TCO is one of the costly components of DSSC [22] but also offers the optically loses for the incident photons to the bottom cell coming through the top cell. Removal of TCO layers as much as possible from the tandem DSSC architecture is expected to be one of the amicable solutions for such issues. We have shown that even after removal of costly TCO components, observed PCE is only slightly lower as compared to their TCO based DSSC counterparts in these cases of both of the Iodine as well as cobalt complex based redox electrolytes [23-24]. Utilization of TCO-less DSSC The utilization of TCO-less DSSCs as bottom electrode in tandem cell have also to enhance the photon harvesting ultimately promoting the PCE [25]. Present article deals with fabrication and characterization of mechanically stacked and series connected TCO-less tandem DSSCs in different device architectures aiming towards optimization of the intermediate layers and its implication on the photon harvesting with various complementary dye combinations.

2. Materials and Methods

The F-doped SnO_2 glass was purchased from OPV tech. All solvents were purchased and used without further processing. For catalytic action a very thin layer platinum (Pt) were coated on the TCO layer by CFS-4EP-LL, (Shibaura Mechatronics, Japan). Sensitizing dyes 2-Cyano-3-[4-[4-(2,2-diphenyl-ethenyl)phenyl]-1,2,3,3a,4,8b-hexahydrocyclopent[b]indol-7-yl]-2-propenoic acid (D131) and Di-tetrabutylammonium-cis-bis(isothiocyanato)bis(2,2'-bipyridyl-4,4'-dicarboxylato)ruthenium(II) (N719) were purchased from Mitsubishi Paper mills Ltd. and Solaronix, respectively. The electronic absorption spectra of dyes used in the present work has been measured using UV-visible spectrophotometer (JASCO, V550). The photovoltaic characterization were measured with solar simulator (CEP-2000 Bunko Keiki BSO-X150LC) and the photocurrent action spectra were performed with constant photon flux of 1 x 10^{16} photons per cm^2 at each wavelength in DC mode. The IV performance were measured under AM1.5 condition with 100 mW/cm^2 solar irradiation by placing of $0.2025cm^2$ black metal mask.

TCO based as well as back contact TCO-less bottom cells were fabricated as per our earlier publication [15]. In order compare the performance of TCO-less tandem DSSC, conventional four TCO-based tandem DSSC were also fabricated. Device architecture of conventional TCO-based tandem DSSC is shown in the figure 1. Two types of conventional TCO-based tandem DSSC (T and T2) were fabricated depending on utilization of sensitizing dyes used in the top and bottom cells. T-tandem DSSC consists of visible photon harvesting sensitizing dyes D-131 and N-719 dyes in the top and bottom cells. The photoanode sensitized with 0.5 mM D131 (1:1 (v/v), acetonitrile: t-butyl alcohol) was assembled with 1 nm transparent Pt coated FTO glass and constitutes the top cell. Mesoporous TiO_2 layer of the bottom cell photoanode was constructed by 15-20 nm particle size of semi-transparent D/SP paste and sensitized by 0.3 mM N719 (1:1 (v/v), acetonitrile: t-butyl alcohol) was assembled with 60 nm Pt sputtered FTO glass. The electrolyte utilized in top cell and bottom cell includes I_2 (50 mM), LiI (500 mM), 4-tert-butyl pyridine (580 mM) and 1-ethyl-3methylimidazolium iodide (600 mM) in dehydrated acetonitrile. On the other hand T2-tandem DSSC was fabricated by utilizing visible light harvesting N-719 dye in the top cell while NIR photon harvesting dye PC-25 in the bottom cell. The top cell photoanode sensitized with N719 (3μm) was assembled relatively thicker (14 μm) bottom cell photoanode. Mesoporous TiO_2 paste with larger particle size (D37) favouring the better dye absorption of bulky PC25 were employed in bottom cell photoanode. Electrolyte involved were Top cell: I_2 0.05M, LiI 0.1M, 4-tBuPy 0.5M and 1,2DPII 0.6M and bottom cell: I_2 0.05M, LiI 1M and 1,2DPII 0.6M in dehydrated acetonitrile.

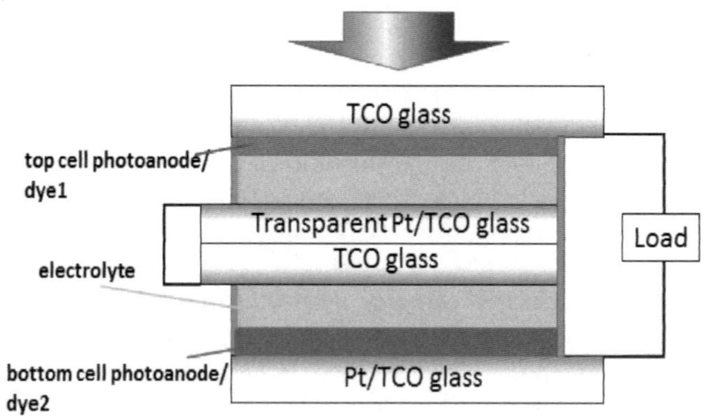

Figure 1. Mechanically stacked conventional tandem DSSCs (T/T2 tandem DSSCs) architecture showing superfluous intermediate TCO layers.

TCO-less tandem DSSCs were fabricated by replacing the TCO-based bottom cell of the tandem DSSC with the back contact TCO-less bottom cell. Device architectures adopted in the TCO-less tandem DSSC fabrication has been schematically shown in the figure 2. Two types of TCO-less tandem DSSCs (T1 and T3) which are analogous to their corresponding conventional TCO based tandem DSSC counterparts. TCO-less back contact DSSCs consisting of glass or plastic film/electrolyte-dye/TiO_2/protected SUS metal mesh/PTFE-spacer/Pt-FTO glass was fabricated as per our earlier publications [24]. The fabrication process basically involves the screen printing of mesoporous TiO_2 paste (D/SP) on protected SUS mesh followed by baking at 450°C for 30 minutes. Sensitizing dyes D131 and N719 were used in the top and bottom cells, respectively of the T1 TCO-less tandem DSSC. On the other hand, T3 TCO-less tandem DSSC includes dyes N719 and PC25 in the top and bottom cells, respectively. Electrolyte composition for respective dyes were the same as used in the conventional tandem DSSCs (T and T2). It should be noted that TCO-less tandem DSSC (T1) utilizes three TCO layers while TCO-less tandem DSSC architecture T3 not only uses only two TCO layers but also the flexible and more transparent ITO/PET film as intermediate layer.

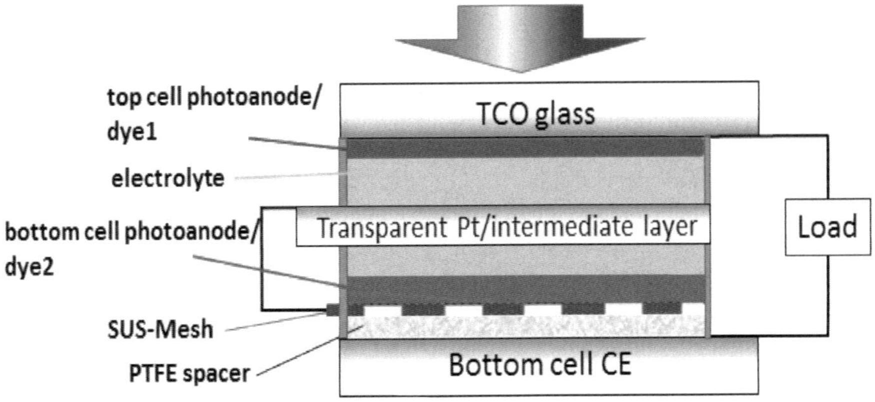

Figure 2. Mechanically stacked TCO-less tandem DSSCs architecture (a) T1 TCO-less tandem DSSCs: FTO/TiO_2/D131/Pt-FTO/N719/TiO_2/SUS-mesh/PTFE/Pt-FTO (b) T3 TCO-less tandem DSSCs: FTO /TiO_2/N719/Pt-ITO-PET/PC25/TiO_2/SUS-mesh /PTFE/Pt-Ti foil.

3. Results and Discussion:

Tandem DSSCs take the advantages of possibility for wide wavelength photon harvesting by the independent and judicious selection of sensitizing dyes having complementary electronic absorption. This offers the possibility of not only enhancing Jsc by wide wavelength photon harvesting but also enhancement in the Voc by connecting the top and bottom cells in series. In the case of series connected tandem solar cells non-complimentary light absorption by top and bottom cells, although leads to enhanced Voc but final out external PCE is either unchanged are slightly improved [26-27]. Taking these points in to consideration we have utilized sensitizing dyes D131, N719 and PC25 having complementary electron absorption spectra which has been shown in the figure 3 along with their molecular structure [15, 25].

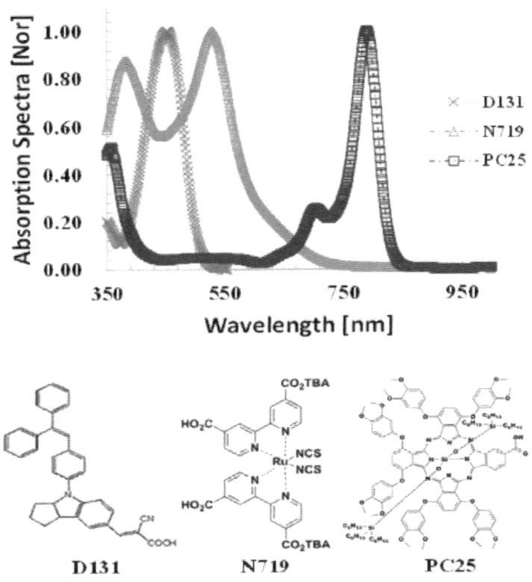

Figure 3. Solution-state electronic absorption spectra of sensitizing dyes D131, N719 and PC25 along with their molecular structure.

It can be clearly seen that dye D131 has a very strong light absorption between 350-550 nm regions where N719 exhibits diminished absorption. At the same time, PC25 exhibits intense light absorption in the NIR 750-900 nm where N719 is unable to absorb any light in this region. This justifies the selection of D131 with N719 or N719 with PC25 for the series connected mechanically stacked tandem DSSC fabrication. As a proof of concept and device functioning of mechanically stacked and series connected tandem DSSCs, a set of sensitizers (D131, N719) having complementary light absorption mainly in the visible region were taken in to consideration. First of all, a conventional tandem DSSCs using four TCO glasses with intermediate TCO layer (T tandem DSSCs) in device architecture as shown in the figure 1 was fabricated by mechanically stacking the top cell and bottom cell.

In this, incident photons have to travel through the dye-adsorbed mesoporous TiO_2 layer of top cell to the bottom cell. During this, photons of high energy are absorbed by the dye stained top cell and remaining un-utilized low energy photons are transmitted to the bottom cell via intermediate layer for the photon harvesting. Since best tandem performance can only be obtained by the proper matching of Jsc from top and bottom cell, it is necessary to control the thickness top cell photoanode and has been found to be of 1 μm in the case top cell using the D131 dye [15]. Photovoltaic performance of the various tandem DSSCs fabricated in this present work has been shown in the figure 4 along with their photovoltaic parameters shown in the table 1. Under optimized thickness of D131 sensitized top cell (1 μm) and N719 sensitzed bottom cell (14 μm), this T tandem DSSC exhibits a short-circuit current density (Jsc) of 6.96 mA/cm^2, open circuit voltage (Voc) of 1.40 V and a fill factor (FF) of 0.65 leading to the external PCE of 6.96 % under simulated solar irradiation. In this T tandem DSSC, a perusal of photovoltaic parameters of individual top cell (PCE 3.53 %, Jsc 7.17 mA/cm^2, FF 0.67, Voc 0.73 V) and bottom cell (PCE 2.89 %, Jsc 6.07 mA/cm^2, FF 0.71 V, Voc 0.67 V) clearly corroborates not only the series connected tandem formation due to summation of top and bottom cell Voc (1.40 V) but also the enhanced external PCE of 6.96 % due to complementary photon harvesting by the respective dyes.

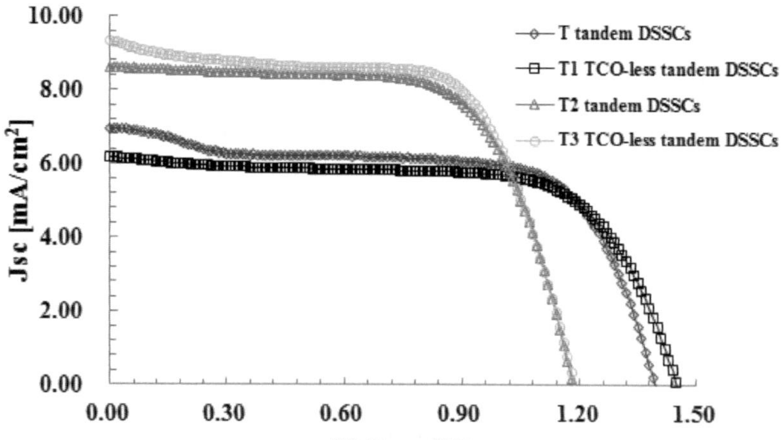

Figure 4. Photovoltaic characteristics of various type of tandem DSSCs after simulated solar irradiation fabricated for present investigation.

Conventionally fabricated T-DSSCs have mechanical rigidity owing to the presence of four TCO glasses which offers the handling and transportation purposes. Apart from this, intermediate TCO layers are not only liable towards reflection and transmission induced photon losses but also imposes addition cost burden. This indicates that suitable countermeasures are needed to adopt the tandem device architectures which are not only efficient in terms of optical management but also cost effective. Amongst available architectural approaches, utilization of TCO-less back contact DSSCs established themselves as one of the cost effective and optically efficient device architectures. Keeping these points in mind, TCO-less tandem DSSC (T1 TCO-less DSSC) was fabricated by stacking a TCO-less DSSCs as bottom cell with the TCO-less top cell DSSCs in figure 2 (device a). Superiority of this fabricated T1 TCO-less tandem DSSCs lies in its economical architecture in which one of the four TCO layers from T tandem DSSC (figure 2a) was removed. This T1 TCO-less tandem DSSCs bearing economical architecture also offers advantageous impact in terms improved light transmission from top cell and intermediate TCO layers reaching to bottom cell is shown in the figure 5. It clearly demonstrates the preference and optical advantage available at bottom cell photoanode by added up 20% photons with respect to conventional T tandem DSSCs [25].

Table 1. Photovoltaic parameters for various tandem-DSSC fabricated in the present work.

Type of tandem DSSCs	PCE (%)	FF	V_{oc} (V)	J_{sc} (mA/cm^2
T Tandem	6.28	0.65	1.40	6.96
T1 TCO-less	6.06	0.68	1.45	6.17
T2 Tandem	6.88	0.68	1.18	8.61
T3 TCO-less	7.19	0.65	1.19	9.34

In order to attain the maximum photovoltaic performance by this T1 TCO-less tandem DSSCs, it is necessary to optimize the performances of top and bottom sub cells especially to match the output Jsc promoting the enhanced FF. Optimization of thickness and dye combinations in top and bottom cells aiming to towards enhancing the PCE of conventional tandem DSSCs have also been systematically performed by Yamaguchi et al [8]. In this work authors clearly demonstrated that mismatch of Jsc of top and bottom cells leads to highly hampered PCE especially due to poor FF. In this work, performances of individual top and bottom cells were first optimized for comparable Jsc which were found to be Jsc 6.12 mA/cm^2, Voc 0.76 V, FF 0.66, PCE 3.06 %) and Jsc 5.79 mA/cm^2, Voc 0.68 V, FF 0.72, efficiency 2.83 % for TCO based top cell and TCO-less bottom cells, respectively. Mechanical stacking and series connection of these optimized top and bottom cells, this T1 TCO-less tandem DSSC, exhibits a Jsc of 6.17 mA/cm^2, Voc of 1.45 V and FF of 0.68, leading to external PCE of 6.06 % under simulated solar irradiation (figure 4 and table 1). This clearly indicates that even after removal of one of TCO layer from the conventional tandem DSSC using four TCO layers final external PCE is not being much affected.

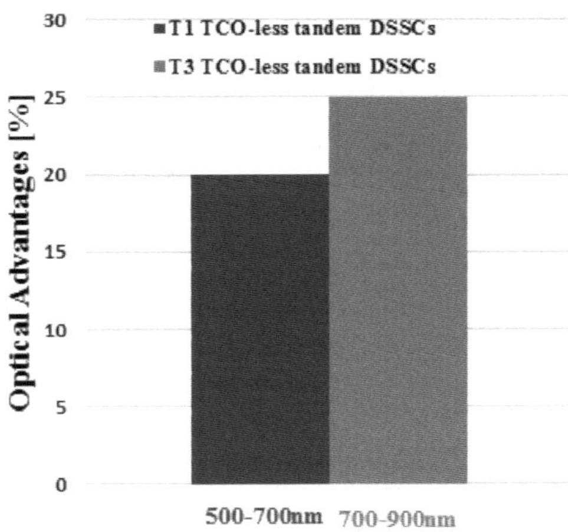

Figure 5. Optical advantages associated with (**a**) T1 TCO-less tandem DSSCs (b) T3 TCO-less tandem DSSCs with respect to respective dye stained conventional T/T2 tandem DSSCs.

Solar spectrum is consisted of photons encompassing from the ultraviolet to far IR wavelength region, therefore a panchromatic photon harvesting in wide wavelength region is inevitable for the enhanced photovoltaic performance of the solar cells. After verification of functioning and tandem formation using visible photon harvesting D131 and N719, efforts have been directed to attain panchromatic photon harvesting by judicial selection of dyes having complementary light absorption. To achieve this goal, series connected and mechanically stacked conventional tandem DSSC (T2 tandem DSSC, figure 1) was fabricated using visible photon harvesting N719 dye in the top cell while complementary NIR light absorbing Si-Phthalocyanine dye (PC25) in the bottom cells. First of all optimization of top cell and bottom cell photoanode thicknesses along with electrolyte composition were conducted in order match the maximum possible Jsc as per our earlier publication before the T2 tandem DSSC fabrication [25]. The photovoltaic performances for optimized top and bottom cells were found to be Jsc 8.40 mA/cm^2, Voc 0.40 V, FF 0.61, PCE 2.06 % (bottom cell) and Jsc 8.72 mA/cm^2, Voc 0.78 V, FF 0.70, PCE 4.77 % (top cell) after simulated solar irradiation. Thus it can be seen that wide photon harvesting due to utilization of visible and NIR dyes has led to the possibility of matching the higher Jsc. Mechanically stacked series connected T2 tandem DSSC fabricated using these optimized TCO based top and bottom cell led to attain a Jsc of 8.61 mA/cm^2, Voc of 1.18 V, FF of 0.68, resulting in to external PCE of 6.88% (figure 4 and table 1) which is higher as compared to its T tandem DSSC counterpart.

Wearable and flexible DSSCs have shown the prominence in utilization of plastic based TCO layers. ITO-PET film exhibits light weight is suitable for the large scale roll to roll fabrication of printed photovoltaic devices. At the same time, Titanium (Ti) foil is has been emerged as cost-effective and flexible alternative to TCO glass along with higher electrical conductivity [28]. Keeping these points in to consideration, further refinement in the T1 TCO-less tandem DSSCs architecture has been carried out in terms of proposal of T3 TCO-less tandem DSSCs as shown in the figure 2 (b). In this T3 TCO-less tandem DSSC, not only flexible intermediate ITO-PET has been used but also flexible Pt-coated Ti-foil has been used as the counter electrode of the TCO-less bottom cell. Utilization of this ultrathin Pt-coated ITO-PET as intermediate layer not only offers the possibility towards the realization of flexible DSSC but also it offers the optical transmission advantage where about 25% more photons are available at the bottom cell photoanode as shown in the figure 5. Top and bottom cell dyes, optimized thicknesses and electrolyte compositions were the same as used in the conventional T2 tandem DSSC. The photovoltaic performances shown by respective sub cells were found to be Jsc 8.13 mA/cm^2, Voc 0.40 V, FF 0.62, PCE 2.04 % and Jsc 9.65 mA/cm^2, Voc 0.75 V, FF 0.67, PCE 4.82 % for the bottom and top cells, respectively. Under simulated solar irradiation this T3 TCO-less tandem DSSC exhibits a photovoltaic performance of 7.19 % (Jsc 9.34 mA/cm^2, Voc 1.19 V, FF 0.65) which is higher than its conventional T2 tandem DSSC counterpart (figure 4 table 1). Therefore, by judicious selection of device architecture and dye combinations and top and bottom cells, it is not only possible to remove two costly TCO components but also the enhanced PCE for the T3 TCO-less tandem DSSCs.

4. Conclusions

In summary, we have successfully demonstrated the fabrication and characterization of light efficient and cost-effective TCO-less tandem DSSCs utilizing visible wavelength region photon harvesting sensitizing dyes D131 and N719. Further modification in the TCO-less tandem architecture was succeeded with fabrication of flexible TCO-less bottom cell and optically more transparent ITO-PET employing Pt-coated Ti foil as counter electrode and intermediate layer, respectively. Introduction of panchromatic photon harvesting by utilizing visible dye N719 in top cell and Si-phthalocyanine dye PC25 in bottom cell and their implementation to fabricate the T3 tandem TCO-less DSSC led to the best external power conversion efficiency of 7.19 % under simulated solar irradiation. In spite of the removal of two costly TCO components out of four in conventional series connected tandem DSSC, there was no any adverse effect on the observed external PCE. Efforts are still needed to design and develop novel NIR dyes with relatively higher Voc in order to attain further enhancement in the PCE.

Acknowledgments
This work was supported by New Energy and Industrial Technology Development Organization (NEDO).

References

[1] Gratzel M, 2006 Prog. Photovolt: Res. Appl. **14** 429.
[2] Kakiage K, Aoyama Y, Yano T, Oya K, Fujisawa J I, Hanaya M, 2015 *Chem Comm.* **51** 15894.
[3] Green A N M, Palomares E, Haque S A, Kroon J M, Durrant J R, 2005 *J. Phys. Chem. C* **109** 12525.
[4] Ehret A, Stuhl L, Spitler M T, 2001 *J. Phys. Chem. B* **105** 9960.
[5] Inakazu F, Noma Y, Ogomi Y, Hayase S, 2008 *Appl. Phys. Lett.* **93** 093304.
[6] He J, Lindstrom H, Hagfeldt A, Lindquist S E, 2000 *Sol. Energy Mater. Sol. Cells* **62** 265.
[7] Wataru K, Ayumi S, Takayuki K, Yuji W, Shozo Y, 2004 *Photochem. Photobiol. A: Chemistry,* **164** 33.
[8] Yamaguchi T, Uchida Y, Agatsuma S, Arakawa H, 2009 *Sol. Energy Mater. Sol. Cells* **93** 733.
[9] Liska P, Thampi K R, Grätzel M, Brémaud D, Rudmann D, Upadhyaya H M, Tiwari A N, 2006 *Appl. Phys. Lett.* **88** 203103.
[10] Wang W L, Lin H, Zhang J, Li X, Yamada A, Konagai M, Li J B, 2010 *Sol. Energy Mater. Sol. Cells* **94** 1753.
[11] Greg D B, Paul G H, Seung-Hyun A L, Neal M A, Janine M, Thomas E M, Paul L, Zakheeruddin S M, Grätzel M, Anita H B, Martin A G, 2011 *J. Phys. Chem. Lett.* **2** 581.
[12] Sobuś J and Ziółek M, 2014 *Phys. Chem. Chem. Phys.***16** 14116.
[13] Ingmar B, Martin K, Felix E, Jaehyung H, Peter E, Anders H, Jürgen W, Neil P, 2009 *Sol. Energy Mater. Sol. Cells* **93** 1896.
[14] Powar S, Bhargava R, Daeneke T, Gotz G, Bauerle P, Geiger T, Kuster S, Nuesch F A, Spiccia L, Bach U, 2015 *Electrochim. Acta* **182** 458.
[15] Baranwal A K, Shiki T, Ogomi Y, Pandey S S, Ma T, Hayase S, 2014 *RSC Adv.* **4** 47735.
[16] Loper P, Moon S J, Nicolas S M, Niesen B, Ledinsky M, Nicolay S, Bailat J, Yum J H, Wolf S D, Ballif C, 2015 *Phys. Chem. Chem. Phys.* **17** 1619.
[17] Ballie C D, Christopher M G, Mailoa J P, Bowring A R, Unger E L, Nguyen W H, Burschka J, Pellet N, Lee J Z, Grätzel M, Noufi R, Buonassisi T, Salleo A, McGehee M D, 2014 *Energy Environ. Sci.* **8** 956.
[18] Xiong D, Chen W, 2012 *Front. Optoelectron.* **5** 371.
[19] Nattestad A, Mozer A J, Fischer M K R, Cheng Y B, Mishra A, Bäuerle P, Bach U, 2010 *Nature. Mater.* **9**, 31.
[20] Seo H, Hashimoto S, Ichida D, Itagaki N, Koga K, 2015 *Electrochim. Acta* **179** 206.
[21] Murayama M, Mori T, 2008 *Thin Solid Films* **516** 2716.
[22] Kroon J M, Bakker N J, Smidt H J P, Liska P, Thampi K R, Wang P, Zakeeruddin S M, Grätzel M, Hinsch A, Hore S, Wurfel U, Sastrawan R, Durrant J R, Palomares E, Pettersson H, Gruszecki T, Walter J, Skupien K, Tulloch G E, 2007, *Prog. Photovolt. Res. Appl.* **15** 1.
[23] Kashiwa Y., Yoshida Y., Hayase S, 2008 *Appl. Phys. Lett.* **92** 033308.
[24] Molla M Z, Mizukoshi N, Furukawa H, Ogomi Y, Pandey S S, Ma T, Hayase S, 2015, *Prog. Photovolt: Res. Appl.* **23** 1100.
[25] Baranwal A K, Fujikawa N, Hayat A, Ogomi Y, Pandey S S, Ma T, Hayase S, 2015 *Appl. Phys. Express.* **8** 102301.
[26] Steinmann V, Kronenberg N M, Lenze M R, Graf S M, Hertel D, Bu¨rckstu¨mmer H, Wu¨rthner F, Meerholz K, 2011 *Appl. Phys. Lett.* **99** 193306.
[27] Wenger S, Seyrling S, Tiwari A N, Grätzel M, 2009 *Appl. Phys. Lett.* **94**, 173508.
[28] Ito S, Ha N L C, Rothenberger G, Liska P, Comte P, Zakeeruddin S M, Pechy P, Nazeeruddin M K, Grätzel M, 2006, *Chem. Commun.* **38** 4004.

Research Trends of Soft Actuators based on Electroactive Polymers and Conducting Polymers

K Kaneto
Department of Biomedical Engineering, Faculty of Engineering, Osaka Institute of Technology, 5-16-1, Ohmiya, Asahi-ku, Osaka 535-8585 Japan

E-mail: keiichi.kaneto@oit.ac.jp

Abstract. Artificial muscles (or soft actuators) based on electroactive polymers (EAPs) are attractive power sources to drive human-like robots in place of electrical motor, because they are quiet, powerful, light weight and compact. Among EAPs for soft actuators, conducting polymers are superior in strain, stress, deformation form and driving voltage compared with the other EAPs. In this paper, the research trends of EAPs and conducting polymers are reviewed by retrieval of the papers and patents. The research activity of EAP actuators showed the maximum around 2010 and somehow declining now days. The reasons for the reducing activity are found to be partly due to problems of conducting polymer actuators for the practical application. The unique characteristics of conducting polymer actuators are mentioned in terms of the basic mechanisms of actuation, creeping, training effect and shape retention under high tensile loads. The issues and limitation of conducting polymer soft actuators are discussed.

1. Introduction

Since the industrial revolution in 18th century, the consumption of natural resources like metals and fossil fuels has been increasing extensively. As the result, the global environment is suffering from the greenhouse effects and depletion of resources. To solve the problems industrial technologies have to change their directions from the mass production and consumption to saving resources and reuse of rare materials. The developments of low energy fabrication techniques and high efficient devices are also urgent issues. Recently, however, clever human beings found that the technology should follow the way of life. Then, the research field turns to direct to life and bio science and technology.

Among research fields of life and bio sciences, the energy sources are intentionally directed to renewable energies, like solar cell, fuel cell, and etc. by the governmental project. Especially, electronic devices using organic materials, that is, organic electronics are being studied intensively. This is because of the low fabrication cost, which results in energy saving, as well as light weight and flexible devices. Most of organic electronics devices are analogy of inorganic solid state devices in the functionality such as transistors (field effect transistor: FET), solar cell and light emitting diode (LED). These organic devices are inferior to the conventional semiconductor devices in element density, response time and efficiency, because of the low carrier mobility, poor crystallinity and purity. However, the novel feature of organic device technology is the fabrication technique, namely, printable devices, such as organic LED color displays, and wearable personal data assistance (PDA).

In organic materials, low molecular weight compounds and functional polymers have been studied for the electronic devices. The low molecular weight compounds like pentacene[1] demonstrated high carrier mobility in FET, because of easy purification and crystallization. On the other hand, the polymers for electronic devices are mostly conducting polymers. The conducting polymers like poly(3-hexylthiophene)[2] are semiconductor and also exhibit large carrier mobility in FET, due to the high crystallinity in region-regular structure. The conducting polymers are also one of electroactive polymers (EAPs). The high conductivity at the oxidized state, inherent flexibility and toughness against mechanical stress are suitable for artificial muscle or soft actuator. The artificial muscle sounds a complicated transducer of chemical energy into the mechanical power. However, the artificial muscles mentioned in this paper are in fact soft actuators, and can't be similar to the complicated natural muscle. Hereafter, we call soft actuator instead of artificial muscle, though the artificial muscle is more attractive.

2. Trend of research activity of electroactive polymers for soft actuator

Presently, ionic polymers (ion exchange membrane) and metal composite (IPMC), conducting polymers, dielectric elastomers, gels (hydrogel and dielectric polymer gels) and carbon nano-tubes (CNTs) are studied as the materials of EAP soft actuators[3]. The trend of research activity on EAP actuators is depicted in figure 1 (a) papers searched by SCOPUS with the keywords of artificial muscle or soft actuator and (b) patents applied in Japan by the keyword of artificial muscle. The term of artificial muscles covers the most of soft actuators. There may be other results by adopting different research items, namely, in title, keyword and abstract. However, the present result statistically suggests the trend of research activity on the EAP actuators. It is found that the study of artificial muscle started around 1990 and showed the maximum activity in 2010, then declining. Taking the fact that patents are disclosed 2 years after the application into consideration, the research activity of soft actuators is apparently decreasing after 2008. The reason can be seen from the decrease of research activity in conducting polymer actuators as clear in figures 1 (a) and (b).

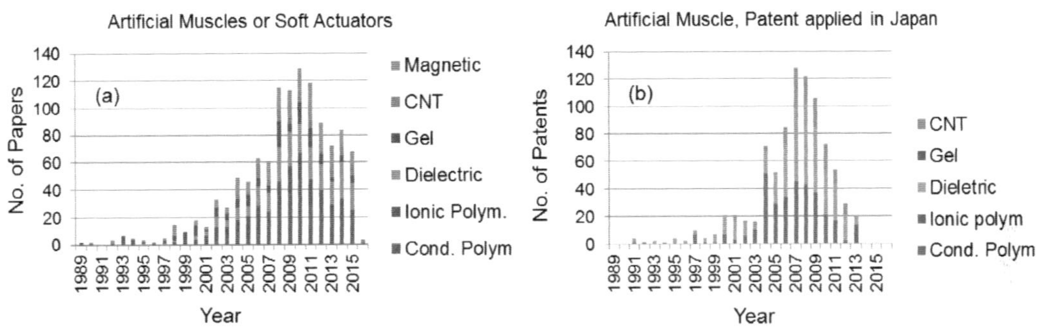

Figure 1. Trends of research activity of soft actuators, (a) the number of papers searched by SCOPUS and (b) the number of patents applied in Japan.

3. Conducting polymers as EAP soft actuators

Conducting polymers have been widely known, since 2000's Nobel Prize of chemistry was awarded to Profs. H. Shirakawa, A. G. MacDiarmid and A.J. Heeger for their invention and study of polyacetylene. Conducting polymers are semiconductor (or insulator) as they are, however, the conductivity increases markedly upon oxidation. The conductivity of oxidized state is as high as metals or more than 10 S/cm. The oxidation is carried out by exposure of conducting polymers to oxidizing agents like bromine and iodine gases or by electrochemical oxidation. During oxidation conducting polymers are swollen by insertion of bulky ion, because of weak inter-chain bonding and fibril structure of the polymer network. On the other hand, the oxidation does not easily occur in solid crystal or inorganic semiconductors, since ions or oxidizing agents hardly penetrate into bulk region. The conductivity, which turns back to those of semiconductors or insulators upon reduction, can be controlled by the applied potential in the electrochemical cell. In the electrochemical reaction, the high conductivity is key factors for the quick response and uniform oxidation in the materials. Consequently, conducting polymers are soft and flexible, mechanically tough and high conductivity, which are suitable to electrically controllable soft actuator. The actuation induced by electrochemical reaction is named as electrochemomechanical deformation (ECMD).

Inventors of soft actuator at the early stage of research using conducting polymers are searched with SCOPUS by the conditions of (artificial muscle)*(conducting polymer). The dimensional change of conducting polymer, polypyrrole, upon electrochemical redox reaction was first reported by Otero et al in 1992[4]. They fabricated bilayer flexible films with polypyrrole and inert films, and demonstrated bending motion due to the expansion and contraction of polypyrrole film by electrochemical redox reaction. They initiated and leaded the research field on conducting polymer actuators until now. Using polyaniline film in 1993, Pei et al[5] reported a bilayer laminate strip actuator, which demonstrated bending motion. Bending micro actuator was reported by Smela et al[6] using conducting polymers in 1993. The *in-situ* measurement of deformation of bilayer polyaniline film and cyclic voltammetry was reported by Takashima et al[7] in 1995 using laser displacement meter. They also succeeded to measure directly the rate of anisotropic elongation of polyaniline film upon electrochemical redox reaction, and discussed the mechanism of actuation[8]. Spinks and Wallace et al[9] discussed detailed mechanism of electrochemical deformation in polypyrrole films in 1995. Baughman[10] reported the conducting polymer actuators as the transducer of electrical energy to mechanical energy as well as energy conversion efficiency.

4. Trend of research activity of conducting polymer soft actuators

Figure 2 shows the number of articles searched by SCOPUS with the condition of actuator (title of paper) and conducting polymer (keyword), and the number of patents by SRPARTNER of US/EU countries with the condition of (artificial muscle + actuator)*(conducting polymer) by the claim. Papers on actuators based on conducting polymers appeared form 1992 and showed the maximum at 2010, then slightly declined in the number. On the other hand, the number of patents applied appeared in 1990 and showed the maximum in 2008 and gradually decreased. It is interesting to note that patents were applied a little earlier than papers published. This indicates that the right of patent is superior to writing papers. Anyways, the research activity for soft actuators of conducting polymers as well as other electroactive polymers seems to be declining now days.

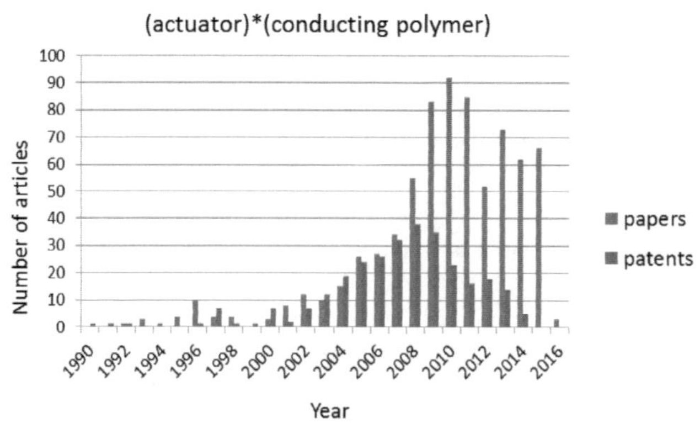

Figure 2. Number of papers and patents on conducting polymer actuators.

From the study of conducting polymer actuators during the last two decades, the attractive points are (1) low operating voltage, (2) large strain, (3) large contraction force (stress), (4) quick response and (5) easy preparation (synthesize) in the optimized circumstance, in comparison with other EAPs. However, we have faced several issues to be overcome before the commercialization of conducting polymer actuators. We will discuss the issue and trend of activity on conducting polymer actuators.

Figure 3 depicts the number of papers of actuators using various conducting polymers, namely, polypyrrole (PPy), polyaniline (PANi), polythiophene, poly(3,4-ethylenedioxythiophene) (PEDOT) and carbon nanotube (CNT). The number of papers is searched by the condition of (actuators)*(PPy) (or the name of conducting polymers) in SCOPUS. The number of paper related to PPy is 313, which is nearly half of the total number of conducting polymer actuators. The next one is PANi of 116 and PEDOT of 57. That of CNT is 52 and polythiophene is 11. The others are negligible. It is noted that the study of PPy is declining and determines the trend of research activity. The research activities for PANi, PEDOT and CNT are miner, however, nearly constant in the recent decade, as seen in figure 3.

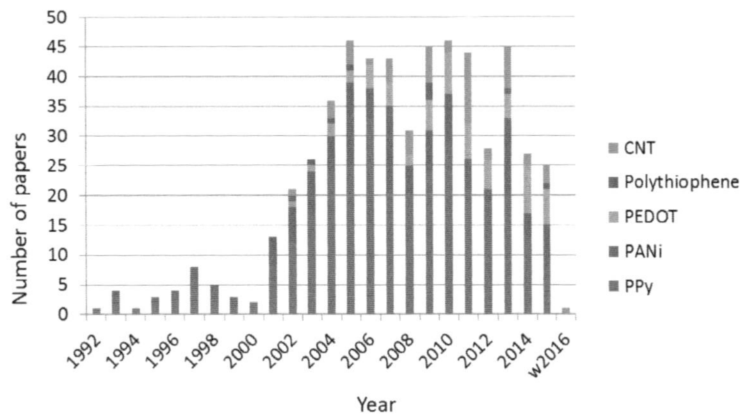

Figure 3. Number of papers of various conducting polymers studied for actuators.

5. Various conducting polymers for soft actuators

Chemical structures of PPy, PANi and PEDOT are shown in figure 4 (a), (b) and (c), respectively. These polymers are aromatic structure in the unit molecule, hence, generally they are stable polymers compared with linear polymer like polyacetylene. These are mostly studied for soft actuators.

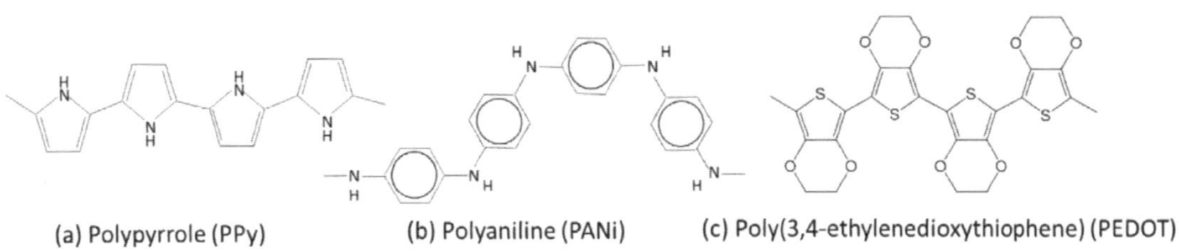

Figure 4. Chemical structures of PPy, PANi and PEDOT

5.1. Polypyrrole

As is noted that the PPy is the most studied conducting polymer for actuators, the reason is that PPy is easily electrodeposited on appropriate metal electrode, and is obtained as high conductive and tough films. Namely, PPy can be prepared in aqueous or organic electrolytes in either form of high density compact film or porous gel like film. Furthermore, the PPy film exhibits large electrochemical strain[11] of 39% and force[12] of 49 MPa at the optimized condition in actuation, and long cycle life[13] in addition. It is noted, however, that preparation of PPy by chemical polymerization in the form of tough and flexible film has not been known, so far. Also PPy obtained by chemical methods are usually insoluble either in aqueous or organic solutions.

5.2. Polyaniline

The major reason for PANi being studied next to the PPy is that PANi is prepared chemically in bulk by the oxidative polymerization in aqueous acid solution of aniline. The reduced state of emeraldine base form is soluble in N-Methyl-2-Pyrolidone (NMP) and Dimethyl Sulfoxide (DMSO), that is, easy fabrication of the solution into films, fiber or desired shapes. The PANi is electrochemically active in acid at pH < 3. The acidic electrolyte is inferior for practical actuators, since leakage of acidic solution is harmful. It has been studied intensively to increase pH for safety operation by using high concentration of supporting electrolyte. It has been found that the highest pH region for stable operation[14,15] is pH4. Another demerit of PANi is the strain of 7% at most, which is smaller compared with that of polypyrrole.

5.3. PEDOT/PSS

PEDOT/PSS; (PSS; poly(styrene sulfonate)), is the most intensively studied conducting polymer for variety of application, such as transparent electrode, capacitor and conductor, because of the high stability, conductivity and easy handling. For the actuator[16,17], the electrochemical strain is limited to 2-3 % and the toughness is not good as PPy and PANi. PEDO/PSS films mostly show cathodic expansion[17,18], indicating anion immobilized in the film and cation movement. However, PEDOT is still interested in studying as actuators, because of PEDOT being high stability in electrochemical reaction. Okuzaki et al[19] demonstrated PEDOT/PSS film actuators, in which water content is controlled by joule heating of the film, namely, at high temperature water goes out and the film shrinks.

5.4. Carbon Nano Tube (Bucky film)

The CNT is also interesting material for actuators[20] besides high conductivity and high mechanical strength. The CNT actuators, which are fabricated by triple layered Bucky films, namely, the porous film soaped with electrolyte was sandwiched by Bucky films. The CNT actuators demonstrate the bending motion similar to IPMC actuator[21]. The actuator showing bending motion is simply interesting for the demonstration of moving. However, the usage of bending motion is rather difficult for the practical application, because the bending force is suppressed by the flexibility of film.

6. Basics of electrochemomechanical actuation in conducting polymers

The mechanism of ECMD in conducting polymers is shown in figure 5. Usually, conducting polymers are obtained by either chemical or electrochemical oxidative polymerization of monomers; hence, the as-grown form is oxidized as shown in figure 5 (a). The oxidized state is swollen by the total volume of inserted anions A^-, and stiffened due to the ionic crosslink. The ionic crosslink is formed by the ionic bonding between polarons and anions, resulting in fastening inter-polymer chain bonding. The oxidized state is reduced by the application of negative voltages in electrochemical cell to figure 5 (c) or (e) by either way of figure 5 (b) or (d), depending on the size of anion used for the polymerization. When small anion like *p*-phenol sulfonic acid (PPS)[22] is employed, the anion can be ejected during reduction and the conducting polymer shrunk as shown by figure 5 (c) through (b). Upon oxidation, the conducting polymers expand, following the reversed way from figure 5 (c) to (a) via (b). In this case, move-in ions are anion, then, named as the anion drive.

On the other hand, if large anion like dodecylbenzene sulfonic acid (DBS)[22] is employed in the synthesis, the large anion is immobilized inside of polymer, and the conducting polymer is reduced by inserting cation to neutralize anions as shown in figure 5 (e) via (d). Namely, the conducting polymer is swollen furthermore. In this case move-in ions are cation, then, the actuation is named as the cation drive. With this idea, a linear actuator is fabricated by laminating with anion and cation dive conducting polymer films on an inert stretchable film. When the actuator is operated by applying voltage between two polymers, the actuator will show the linear motion.

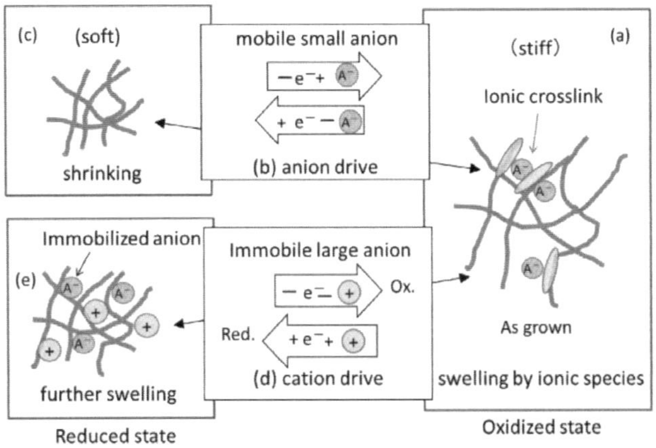

Figure 5. Mechanism of electrochemomechanical actuation in conduction polymers.

The oxidized state is stiffer compared with that of reduced state by about 2-3 times. For example, as shown in figure 6, the Young's modulus (Y) of PPy/DBS at oxidized state is 0.3 GPa and that of reduced state is 0.11 GPa[3]. The curves are obtained by applying tensile stress to a film, and measure the change of film length. The stress (σ) is plotted against strain (ε) as shown in figure 6, namely, stress-strain curve. The Young's modulus (Y) is obtained from the gradient of the curve, $\sigma = \varepsilon\ Y$, at the linear portion. The high stiffness (large Y) is originated from the delocalization of π-electrons, *i.e.*, polaronic states and the ionic crosslink. On the other hand, at the reduced state the polaronic state and ionic crosslink disappear, then the PPy comes soft. In this way, it is interesting to note that the rheology of conducting polymers can be controlled by electrochemical oxidation and reduction.

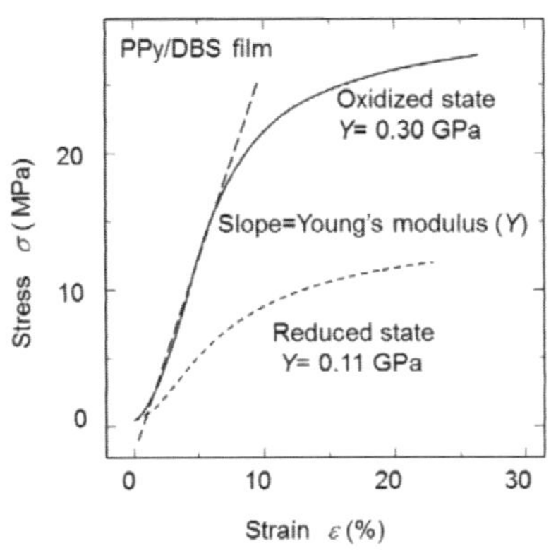

Figure 6. Static stress-strain curves of PPy/DBS films at oxidized and reduced states.

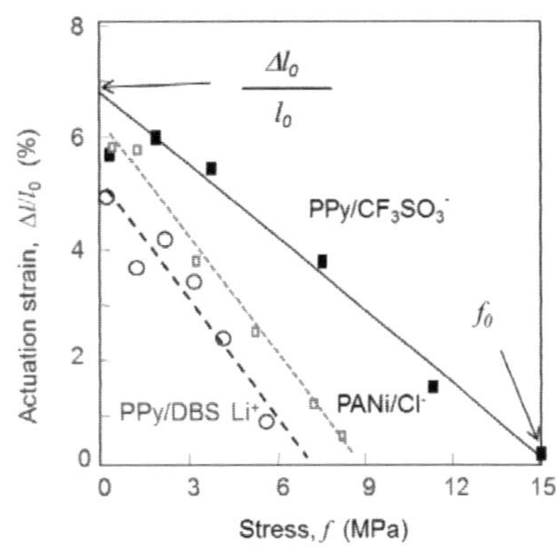

Figure 7. Actuation Strain-Stress curves in of PPy and PANi films.

$$\frac{\Delta l}{l_0} = \frac{\Delta l_0}{l_0} - \frac{f}{Y} \qquad (1)$$

By the electrochemical reaction, the conducting polymer films changes in the dimension, and the length of films is measured and defined by the change of length (Δl) (or stroke). The actuation strain is given by dividing Δl by the original length (l_0) of the film, i.e., $\Delta l/ l_0$ in %. The ECMD strain ($\Delta l/ l_0$) depends on the size of ions and degree of oxidation[3]. The $\Delta l/l_0$ also depends on the tensile loads, namely, $\Delta l/l_0$ decreases with increasing the tensile loads (stress; f) as shown in figure 7 by the typical strain-stress curves. The actuation stress (namely, contraction force) strictly depends on the Young's modulus (or stiffness of films), as shown by empirical equation (1), where Δl_0 is the maximum stroke at non tensile load. The f_0 is the maximum contraction force, which is obtained by letting $\Delta l/l_0 = 0$ in equation (1) and shown in figure7. Consequently, Y is obtained from the actuation strain-stress curve by $Y = f_0/(\Delta l_0/l_0)$. It is interesting to note that from the curve of PPy/DBS in figure 7 the Y is estimated to be 0.14 GPa (f_0 = 7MPa, $\Delta l_0/l_0$ = 0.05), which approximately coincides with the average Young's modulus of 0.11 and 0.3 GPa as shown in figure 6.

The result indicates that the empirical equation (1) is useful to discuss the ECMD and how to improve the actuator performance, when the materials are given. The Δl_0 and Y seem to be conflicting parameters, namely, the stiff material is hard and shows small deformation. The issues to improve the actuator performance are to prepare materials with large Δl_0 and Y.

7. Unique characteristic of conducting polymer actuators based on creeping

Creeping is a phenomenon that is observed as the irreversible and uncontrollable elongation, when the tensile load stress goes over the elastic or Hook's law region. In soft actuators the creeping is serious problem for the precise positioning[23-25]. The creeping is thought to be resulted from (1) uniaxial alignment of polymer chains, (2) slipping of polymer chains and (3) partial breaking of polymer chains before the breakdown of film. Figure 8 shows an observation of creeping in PANi film under tensile load stress. The short period waves with the stroke of ca. 0.5 mm were due to ECMD, and the shifting of elongation resulted from creeping, as seen stepwise increase of elongation at 1 and 3 MPa. Creeping is significantly enhanced[24] during actuation under tensile stress as shown in figure 8. However, it is noted that the creeping was recovered by removing the tensile loads and repeat of electrochemical cycle. It was also observed that the shape of elongated film is retained by the quitting of electrochemical cycle, indicating the shape retention effect. Furthermore, the stroke of ECMD was slightly increased by electrochemical cycles after the removal of tensile loads. This phenomenon could be named as the training effect.

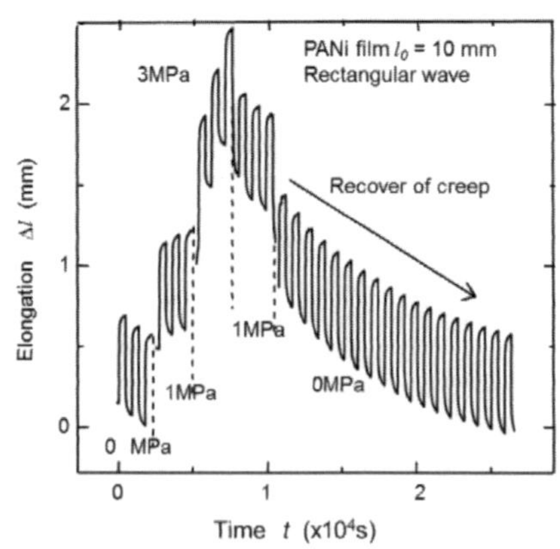

Figure 8. Creeping in PANi actuator under tensile loads of 1 and 3 MPa, and the recovery.

The shape retention and recovery of creeping are unique and interesting phenomena in conducting polymer actuators. These effects result from the conformation change and uniaxial stretching of polymer chains during creeping[23,24]. The mechanism of shape retention is considered to be due to that the tensile stress stretches polymer chains, which is fixed by ionic crosslink. The fixing of aligned conformation with ionic crosslink is recovered to original shape by thermal relaxation of the polymer chain and by the number of electrochemical cycles. The training effect is supposed to originate from the disentanglement of polymer chains and enhancement of electrochemical activity.

8. Epilogue and conclusion

Electronics is a device technology, which controls the movement of electrons in solid materials like semiconductors, and utilizes the resulting optical and electrical functionality for electronic devices. For example, transistors, photodiodes, solar cells, electroluminescence (EL) diodes and etc. are the typical electronics devices. The electronics has been developed since the discovery of transistors by J. Bardeen, W. B. Shockley and W. H. Brattain in 1947. They received Nobel Prize in 1956, 10 years after the discovery.

Iontronics is a new technology, which controls either electrons or ions in solid and/or fluid materials, and utilizes the optical and electrical functionality. For instance, batteries, electrochemical displays, dye sensitized solar cell, electrochemomechanical actuators and etc. are typical iontronic devices. Of course these have been known as electrochemical devices, however, have not been called as iontronics. The organic iontronics is aiming novel devices such as biosensors, enzyme sensors and artificial muscles, in which ions play the key role of the functions. These devices are unique and cannot be realized by the technology of electronics solely.

The high conductivity and flexibility of conducting polymers are important factors for iontronics, because the electrochemical activity depends on the conductivity and the porosity, that is, weak bonding between atoms or molecules. Insulators or rigid metals may not be used in the iontronics. It should be remembered that the conducting polymers are also excellent materials for electronics, like flexible transistors, solar cells and EL diodes.

Artificial muscle or soft actuator is a future device of the human dream for a long time, to fabricate human friendly robots. Soft actuators have been studied for more than 25 years using EAPs including conducting polymers. However, none of them has been commercialized yet. We hope that the research activity on soft actuators come to be aggressive again to realized human friendly robots and etc.

References

[1] Horowitz G, Peng X.-Z, Fichou D and Garnier F 1992 Role of the semiconductor/insulator interface in the characteristics of p-conjugated-origomer-based thin-film transistors *Synthetic Metals* **51** 419

[2] Kaneto K, Lim W Y, Takashima W, Endo T and Rikukawa M 2000 Alkyl Chain Length Dependence of Field-Effect Mobilities in Regioregular Poly(3-Alkylthiophene) Films *Jpn. J. Appl. Phys.* **39** L872

[3] Asaka K and Okuzaki H 2014 *Soft Actuators -Materials, Modeling, Application, and Future Perspectives-* (Springer)

[4] Otero T F, Angulo E, Rodríguez J and Santamaría C 1992 Electrochemomechanical properties from bilayer; polypyrrole/non-conducting and flexible material –artificial muscle *Journal of Electroanalytical Chemistry* **341** 369

[5] Pei Q, Inganäs O and Lundström I 1993 Bending bilayer strips built from polyaniline for artificial muscle *Smart Materials and Structure*, **2** 1

[6] Smela E, Inganäs O and Lundström I 1993 Conducting polymers as artificial muscles: challenges and possibilities *Journal of Micromechanics and Micoengineering* **3** 203

[7] Takashima W, Kaneto M, Kaneto K and MacDiarmid A G 1995 The electrochemical actuator using electrochemically-deposited poly-aniline film *Synthetic Metals* **71** 2265

[8] Takashima W, Fukui M, Kaneko M and Kaneto K 1995 Electrochemomechanical Deformation of Polyaniline Films *Jpn. J. Appl. Phys.* **34** 3786

[9] Gandhi M R, Murray P, Spinks G M and Wallace G G 1995 Mechanism of Electromechanical

actuation in polypyrrole *Synthetic Metals* **73** 247
[10] Baughman R, 1996 Conducting polymer artificial muscles *Synthetic Metals* **78** 339
[11] Hara S, Zama T, Takashima W and Kaneto K 2005 Free-standing gel-like polypyrrole actuators doped with bis(perfluoroalkylsurfonyl)imide exhibiting extremely large strain *Smart Mater. Struct.* **14** 1501
[12] Zama T, Hara S, Takashima W and Kaneto K 2005 Comparison of Conducting Polymer Actuators Based on Polypyrrole Dope with BF_4^-, PF_6^-, $CF_3SO_3^-$, and ClO_4^- *Bulletin of the Chemical Society of Japan* **78** 50
[13] Madden J, Rinderknecht D, Anquetil P A and Hunter I W 2007 *Sensors and Actuators A: Physical* **133** 210
[14] Takashima W, Nakashima M, Pandey S S and Kaneto K 2004 Enhanced electrochemical activity of polyaniline films towards high pH region: condition of Donnan effect *Electrochimica Acta* **49** 4239
[15] Takashima W, Nakashima M, Hashimoto H, Tominaga K and Kaneto K 2009 Suppression of pH effect on the electrochemical deformation in polyaniline film *Thin Solid Films* **518** 674
[16] Vendeteeg N, Madden P G, Madden J D, Anquetil P A and Hunter I W 2003 Synthesis and characterization of EDOT-based conducting polymer actuators *Proceeding of SPIE* **5054** 349
[17] Kiefer R, Kilmartin P A, Bowmaker G A, Cooney R P and Travas-Sejdic J 2007 Mixed-ion linear actuation of PPy and PEDOT in propylene carbonate-triflate electrolytes *Proceedings of SPIE* **6524** 65240U.
[18] Kaneto K, Takayanagi K, Tominaga K and Takashima W 2012 How to improve electrochemomechanical strain in conducting polymers *Proceeding of SPIE* **8340** 917879
[18] Okuzaki H, Suzuki H and Ito T 2009 Electrically driven PEDOT/PSS actuators *Synthetic Metals* **159** 2233
[20] Baughman R H, Cui C, Zakhidov A A, Iqbal Z, Barisci J N, Spinks G M, Wallace G G, Mazzoldi A, De Rossi D, Rinzler A G, Jaschinski O, Roth S and Kertesz M 1999 Carbon nanotube actuators *Science* **284** 1340
[21] Fukushima T, Asaka K, Kosaka A and Aida T 2005 Fully plastic actuator through layer-by-layer casting with ionic-liquid bucky gel *Angew. Chem. Int. Ed.* **44** 2410
[22] Takashima W, Pandey S S and Kaneto K 2003 Cyclic voltammetric and electrochemical characteristics of freestanding polypyrrole films in dilute media *Thin Solid Films* **438-439** 339
[23] Kaneto K, Hashimoto H, Tominaga K and Takashima W 2011 Shape retention in polyaniline artificial muscles *Jpn. J. Appl. Phys.* **50** 02160
[24] Tominaga K, Hashimoto H, Takashima W and Kaneto K 2011 Training and shape retention in conducting polymer artificial muscles *Smart. Mater. Struct.* **20**124005
[25] Madden J D, Ringerknecht D, Anquetil P A and Hunter I W 2007 Creep and cycle life in polypyrrole actuators *Sensors and Actuators* **A 133** 210

Orientation Characteristics of Non-regiocontrolled Poly(3-hexyl-thiophene) Film by FTM on Various Liquid Substrates

M Pandey[1], S Nagamatsu[2, 3], S S Pandey[1], S Hayase[1,2] and W Takashima[1,2]

[1]Graduate School of Life Science and Systems Engineering, Kyushu Institute of Technology, Wakamatsu, Kitakyushu 808-0196, Japan
[2]Reseach Center for Advanced Eco-fitting Technology, Kyushu Institute of Technology, Wakamatsu, Kitakyushu 808-0196, Japan
[3]Department of Computer Science and Electronics, Kyushu Institute of Technology, 680-4 Kawazu, Iizuka, 820-8502, Japan

E-mail: pandey-manish@edu.life.kyutech.ac.jp / wtakashima@life.kyutech.ac.jp

Abstract. Orientation characteristics of non-regiocontrolled poly(3-hexylthiophene) (NR-P3HT) films prepared by dynamic casting of floating film and transferring method (FTM) has been investigated. The film was first cast on liquid-substrate to obtain as a floating-film followed by its transfer on solid-substrate such as white-glass or Si-wafer in order to evaluate their optoelectronic characteristics. As a possible key-factor to generate the orientation of conjugated polymer in this method we focused on the components of liquid-substrate in this study. The orientation dependence upon various liquid-substrates reveals that dichroic ratio strongly changes with liquid-substrates. Pictures of floating-film show the change in size of floating-parts depending upon the liquid-substrate, representing the expansion length of casting solution upon the viscosity. These findings have indicated that spreading speed of polymer solution and solvent evaporation speed controls the size of floating-film leading to change in the orientation intensity. The multilayer coatings of oriented NR-P3HT films were used for polarized FTIR analysis exhibiting clear dichroism. The obtained dichroic characteristics were well corresponded with in-plane, out-of-plane and non-oriented vibronic modes of P3HT.

1. Introduction

Solution processing is one of the important key-features to prepare organic electronic devices with ease and at larger scales. Such processing feature in organic electronics contributes not only to be cost-effective but also the energy efficient process which is advantageous over the Si-based electronics. However, the organic-semiconductor films prepared with solution process exhibits varied film morphologies due to random mixing and/or alignment of molecules during solid phase condensation. The electronic characteristics like carrier transport drastically changes with the film morphology. Casting procedure to prepare the organic-semiconductor film, therefore, is the most important key-factor to bring out the potential performance of these materials. Origin of this varied film morphology is associated with the intrinsic anisotropy of organic semiconductors in terms of their geometric and electronic structures. Many efforts have been paid to obtain the molecular alignment in casting process in order to control the film morphology. In particular, conjugated polymer (CP) is the most anisotropic organic semiconductor. Thus, the orientation methodology of CP with facile and rapid way is strongly desired in order to deploy the profit of its anisotropy. We have developed a novel casting procedure to promote the orientation of CPs on liquid-substrate, which can be expressed as dynamic casting of floating-film and transferring method (dynamic FTM) [1]. Importance of this procedure lies in the utilization liquid-substrate for casting of a thin film. Liquid-substrate provides the mobile surface to combine under solidifying parts with each other leading to the formation a uniform film. In particular, if we choose a volatile solvent such as chloroform for casing solution, the floating-film solidification occurs immediately during spreading of the solution on liquid-substrate. In this case, floating-film formed under solidification with spreading, leading to compress the film with the motion of liquid-substrate resulting in to an oriented floating-film. This method is quite unique and quick to deposit a centimeter scale of oriented film and it can be coated on various organic surfaces without having interlayer interferences.

Various factors such as volatile speed of solvent, spreading speed of solution, solidification speed of materials will affect on the orientation characteristics in dynamic FTM. In fact, volatile solvents such as tetrahydrofuran and dichloromethane can provide similar oriented films by dynamic FTM at room temperature. Against this, benzene-based solvents such as toluene and chlorobenzene failed to provide oriented floating-film in the same condition. The orientation characteristics vary with the liquid-substrate and the casting temperature, also. Considering these complicated states for obtaining the well-orientated films in dynamic FTM, it is important to proceed the step by step analysis of each casting factor on the orientation characteristics, carefully.

Poly(3-hexylthiophene) (P3HT) is one of the most investigated conjugated polymers pertaining to its semiconductor characteristics. Since the discovery of controlled synthetic method of P3HT to react selectively in the head-to-tail manner via a modified Kumada-Tamao-Corriu cross-coupling [2] with Ni catalyst [3], drastic improvements of transport characteristics have been realized leading to high performance electronic devices such as organic solar-cells [4] and organic thin-film transistors (OTFTs) [5]. Non-regiocontrolled P3HT (NR-P3HT) synthesized with oxidative polymerization using FeCl3 [6, 7], therefore, rather reduced to be investigated as a classical material. NR-P3HT, however, still possesses advantages to polymerize using cheap raw materials with relatively simple methodology and high yield. Recently, we have found that NR-P3HT could be oriented with high dichroic ratio (DR) of 9.3 via dynamic FTM. NR-P3HT is, therefore, regarded as a proper target material for the research of orientation characteristics with dynamic FTM. In this work, we have focused the effect of liquid-substrate on the orientation characteristics in dynamic FTM. For the purpose to vary the viscosity, we choose several mixtures of distilled water (Wt) and ethylene glycol (Eg) as low-viscosity and mixtures of Eg and glycerol (Gl) as high-viscosity liquid-substrates. Orientation characteristics of NR-P3HT films thus prepared have also been clarified by polarized optical measurements.

2. Materials and Methods

NR-P3HT has been oxidative synthesized with $FeCl_3$ catalyst. The detail procedure of polymerization is as follows. 3-hexylthiophene/$CHCl_3$ solution was dropped in $FeCl_3$/$CHCl_3$ slurry at -5°C with stirring. The final ratio of 3-hexylthiophene:$FeCl_3$ in the mixture was set at 1:4. This mixture was reacted with stirring at -5°C for overnight (~12 hs), then quenched with 1M HCl to stop the reaction. The solid part was collected and dissolved in fresh $CHCl_3$. This solution was mixed with NH_4OH aqueous solution. The mixture was vigorously stirred with boiling, then separated aqueous part and added a fresh NH_4OH solution, again. This procedure was repeated for 4 times. After that, the $CHCl_3$ solution was separated and added 0.05M of ethylenediaminetetraacetic acid (EDTA) aqueous solution to remove Iron-based impurities [8]. This mixture was stirred with warming for 1h for purification, repeatedly. The dark orange $CHCl_3$ part was separated, washed with water, then filtrated to remove insoluble part. The fine $CHCl_3$ solution was condensed and precipitated in methanol. Again the obtained solid part was dissolved $CHCl_3$, then precipitated in acetone. The solid part was collected and washed by Soxhlet extraction with hexane in order to separate relatively low-molecular parts [9]. Finally, relative high-molecular NR-P3HT was obtained as dark brown powder. Regioregularity was confirmed to be about 80% by the area ratio of ^{1}H-NMR spectrum peaks at 2.56 (head-to-head coupling) and 2.80 (head-to-tail coupling) ppm as shown in figure 1.

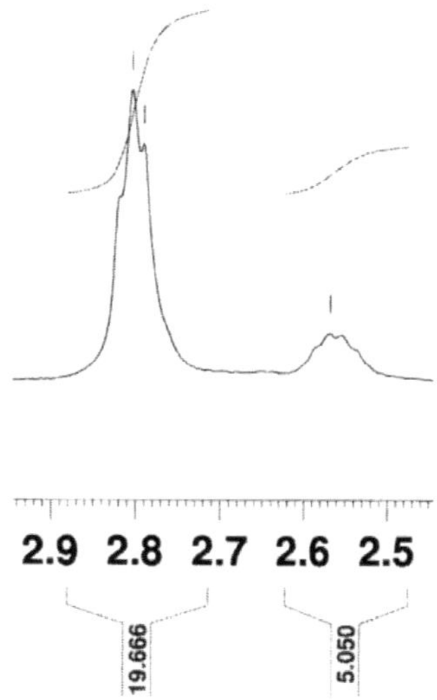

Figure 1. ^{1}H-NMR spectrum at α-methylene region. Area of the peaks are 19.666 at 2.80 ppm (head-to-tail coupling) and 5.050 at 2.56 ppm (head-to-head coupling) regions. This provides 79.6 % as the head-to-tail coupling ratio (regioregularity) of this NR-P3HT.

10 mg of NR-P3HT was typically dissolved in 1g of dehydrated chloroform to obtain 1% (in w/w) solution which was then used to cast thin film by dynamic FTM. Dynamic FTM was conducted by putting a droplet (about 10uL) of solution at the center of the liquid-substrate. The droplet quickly started to spread in all the direction and to solidify as a floating-film, simultaneously. After solidification the floating-film was left for several minutes in order to complete the evaporation of solvent (chloroform). A solid-substrate such as Si-wafer and/or white-glass was put on an oriented part of the floating-film to transfer it on the substrate. The surface of the transferred floating-film was washed with methanol and dried by blowing air to remove dews remaining on the surface. The coated film can be easily confirmed by the naked eyes owing to change in color as compared to non-coated surface.

A single layer of FTM coating provides oriented films of NR-P3HT having the thickness of about 20 nm measured by interference microscopy Nikon Eclipse LV 150. Multiple coatings of oriented NR-P3HT have also been done in order to increase the thickness for spectroscopic characterization. Finally, we could obtain about a 800 nm-thick oriented film by the multiple-coatings. For the purpose to investigate the effect of liquid-substrate viscosity on orientation, a variety of binary mixture of water and ethylene glycol (Wt/Eg) or ethylene glycol and glycerol (Eg/Gl) were used as liquid-substrates. Optical dichroic characteristics were investigated by the polarized absorption spectra obtained through a Glan-Thompson prism with JASCO (V-570) spectrophotometer. Polarized FTIR spectra were also measured by JASCO Fourier transform infra-red spectrometer (FTIR-4100) through polarizer unit. We used DR as the relative height of peak absorption in both the parallel- and orthogonal-spectra at the peak wavelength in the parallel-spectra in this study.

3. Results and Discussion

Figure 2 shows the photographic images for the floating-film deposited on 7 cm-diameter liquid-substrate. It can be seen from figure 2 that size of inner thick and non-uniform part reduces with increasing the viscosity of the liquid-substrate. This suggests that viscosity of liquid-substrate strongly affects the molecular condensation of floating-film during dynamic FTM.

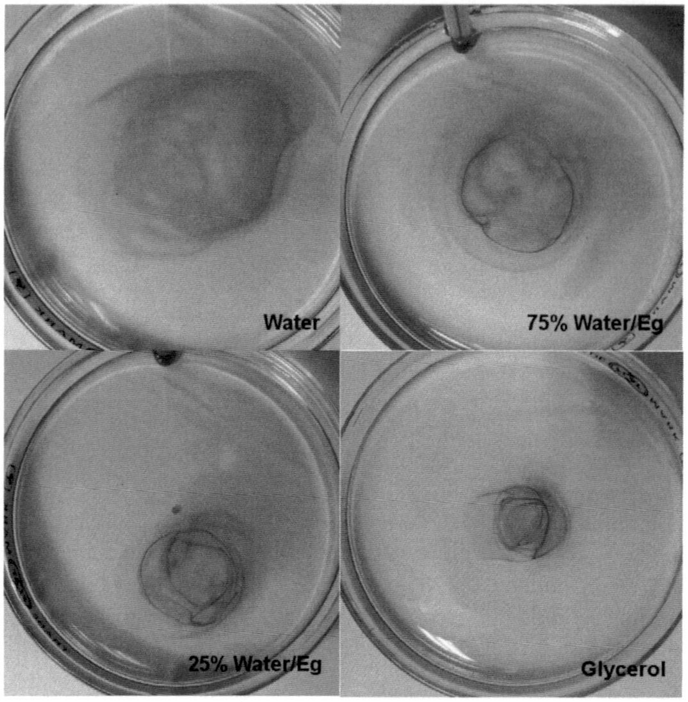

Figure 2. Pictures of floating-film cast in a petri-dish having 7 cm-diameter.

A perusal of thin film formation under dynamic FTM reveals that it is consisted of the three-type parts. The outer (initially formed) dispersive part seems to be very thin (sometimes invisible). The middle one shows fine orientation with good uniformity in thickness. The final inner (central) part, deposited with middle part in continuation is although good in the orientation but with least uniformity. Both of the outer and the middle parts are solidified from the outer to the inner direction. Against this, at the end of solidification the central area filled by remaining solution solidifies from the outer to the center area. The procedure to form the inner part is quite similar to the solidification of drop-cast film on the solid-substrate. The change in the solidification from outer to the inner direction for inner parts suggest that the remaining solution cannot be spread any more, then forms a film with evaporation of solvent, under limiting the motion of solution at remaining narrow area. The immobile characteristics of solution for solidification of the inner parts do not match to the concept for liquid-substrate. Therefore, the most featured part as dynamic FTM is the oriented middle part in floating-film. Therefore, we selected to transfer the middle part of floating-film on solid-substrate as the well-oriented and uniform film for further characterization.

Various binary liquid-substrates were used for obtaining the oriented films. DR obtained by polarized-absorption spectra as shown in the insets in figure 3 depicts the viscosity dependent DR. Viscosities of the liquid substrates used in this work were varied by mixing ratio of water/ethylene glycol (Wt/Eg) or ethylene glycol/glycerol (Gl/Eg). This figure clearly illustrates that the orientation intensity (indexed by DR) strongly depends upon the liquid-substrate. It should be noted that the viscosities of Wt, Eg and Gl were 1, 16 and 1420 mPa.s, respectively. The viscosity of liquid-substrate controls the spreading speed of polymer solution. Therefore, finally inverse of x-axis in figure 3 can be regarded as the spreading speed. By considering the images of floating-film shown in figure 2 along with figure 3 the size of inner part reduces with increasing the spreading speed. These results suggest a detailed mechanism to generate inner part after solidification. Increase in the spreading speed leads to the enhanced coverage area by the outer and the middle parts also. The central part not to be covered by the outer and the middle parts, therefore, reduces in size, which defines the size of the inner parts. This finding of DR dependence on viscosity as discussed above indicates that a sort of optimal matching for both of the spreading speed of solution and the evaporation speed of solvent is expected to promote the orientation of NR-P3HT.

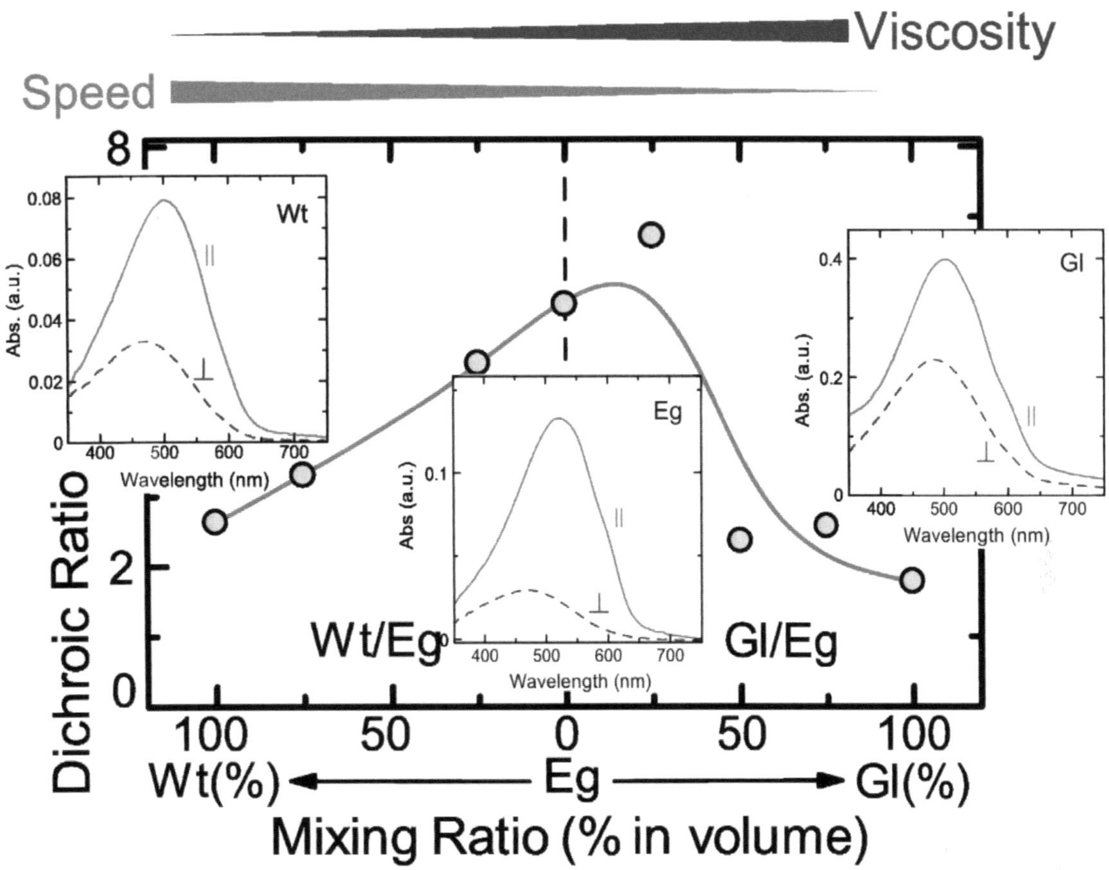

Figure 3. DR as a function of mixing ratio in various binary liquid-substrates. Insets show the polarized absorption spectra of NR-P3HT films cast on Wt, Eg and Gl liquids.

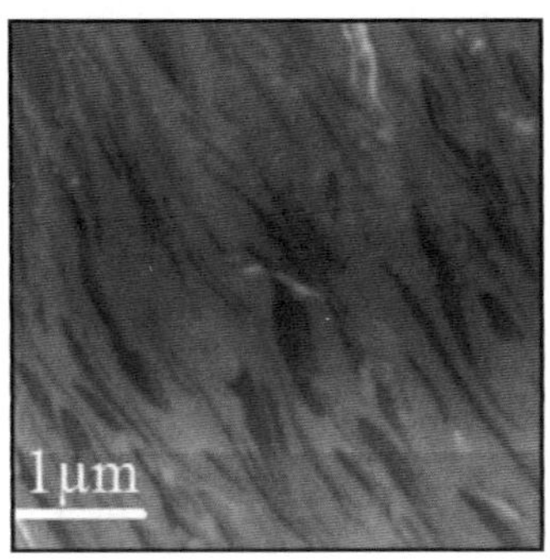

Figure 4. AFM image of oriented NR-P3HT film surface.

Figure 4 shows an AFM image of oriented NR-P3HT film surface. As can be found, NR-P3HT film is consisted of uniaxially-stretched domains. This direction is matched to the parallel direction of the polarized absorption spectra (shown in the insets in figure 3). This indicates that the macromolecules of NR-P3HT lie into the same direction in these stretched domains.

Figure 5 shows the distinctive vibration band of polarized FTIR spectra of an oriented NR-P3HT film. Thanks to the thickness (about 800 nm) prepared by multi-coating with FTM, relatively high-absorptions could be obtained at 1456, 1506 and 1517 cm^{-1} in parallel FTIR spectra. These bands are assigned as the in-plane vibration modes of thiophene ring [10]. Against this, a clear inverse dichroism was found at 825 cm^{-1} with a new band at 835 cm^{-1}. These dual peaks has been assigned as the out-of-plane mode of thiophene ring [10, 11]. Contrary to these bands, non-dichroic absorptions were found at 722 and 1377 cm^{-1}, which are assigned as vibration modes of alkyl side chain of the thiophene substituent. It should be noted that the wide-band area at around 1420-1490 cm^{-1} is known to be consisted of the overlapping both of the thiophene-based in-plane and alky modes [11, 12]. A careful perusal reveals that the band at 1464 cm^{-1} is non-dichroic and the three-peaks at 1419, 1439 and 1456 cm^{-1} shows clear dichroism. In particular, the dichroic ratio at 1419 and 1439 cm^{-1} seems to be similar as those at 1506 and 1517 cm^{-1}. It is also reported that there are three in-plane modes at around 1417-1460 cm^{-1} [11]. These results support that the peaks at 1419 and 1439 as well as 1456 cm^{-1} can be assigned as the in-plane vibration modes of thiophene ring, and 1464 cm^{-1} is belong to alkyl modes, respectively. The finding separable assignments at this complex band can be easily carried out owing to the uniform strong orientation in this thick film.

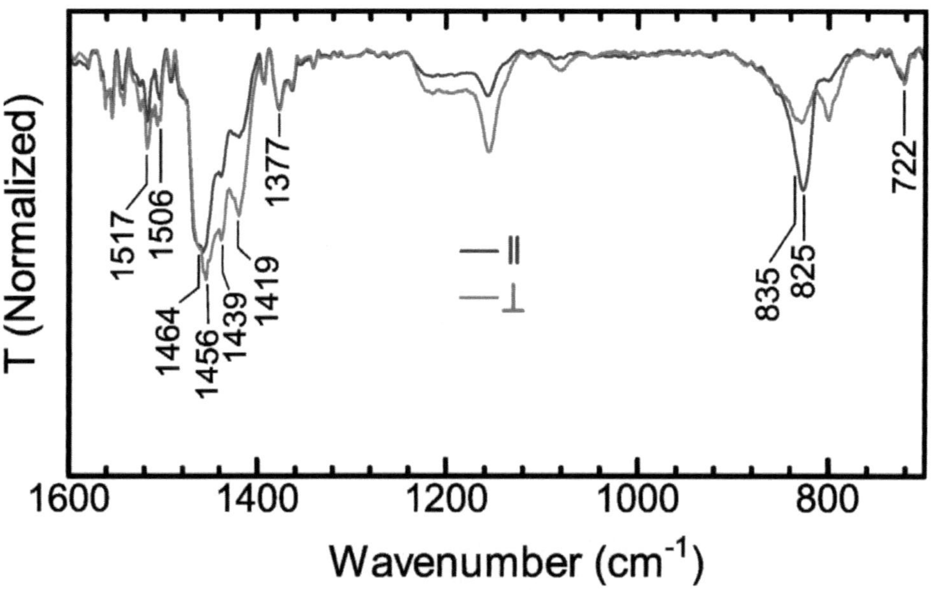

Figure 5. Polazied FTIR spectra of a muli-layerd NR-P3HT film prepared by dynamic FTM.

4. Conclusions

Various binary liquid-substrates were used for obtaining oriented films of NR-P3HT. A correlation between the spreading speed of solution and the solidification speed of floating-film is found to be an important factor to identify the orientation intensity for dynamic FTM. Thanks to the macroscopic multiple parallel-coatings of oriented NR-P3HT by dynamic FTM, various polarized spectroscopic characterization can be easily carried out to various thick oriented films with a simple polarizer.

Acknowledgments

Authors sincerely express their thanks to Japan Society for the Promotion of Science (JSPS) for the financial support by Grant-in-Aid for Scientific Research (C) (Grant Number 15K05989).

Journal of Physics: Conference Series **704** (2016) 012005 doi:10.1088/1742-6596/704/1/012005

References

[1] Dauendorffer A, Pandey RK, Miyajima S, Nagamatsu S, Prakash R, Takashima W, Hayase S and Kaneto K 2013 *Trans. Mat. Res. Soc. Jpn.* **38** 305
[2] Loewe RS, Ewbank PC, Liu J, Zhai L and McCullough RD 2001 *Macromolecules* **34** 4324
[3] Tamao K, Kodama S, Nakajima I, Kumada M, Minato A and Suzuki K 1982 *Tetrahedron* **38** 3347
[4] Yu G, Gao J, Hummelen J C, Wudle F and Heeger A J 1995 *Science* **270** 1789
[5] Bao Z, Dodabalapur A and Lovinger A J 1996 *Appl. Phys. Lett.* **69** 4108
[6] Sugimoto R, Takeda S, Gu HB and Yoshino K 1986 *Chem. Express* **1** 635
[7] Amou S, Haba O, Shirato K, Hayakawa T, Ueda M, Takeuchi K and Asai M 1999 *J. Polym. Sci. Polym. Chem.* **37** 1943
[8] Trznadel M, Pron A, Zagorska M, Chrzaszcz R and Pielchowski J 1998 *Macromolecules* **31** 5051
[9] Pandey SS, Takashima W, Nagamatsu S and Kaneto K 2000 *IEICE T. Electron.* **E83-C** 1088
[10] Garreau S, Louarn G, Buisson JP, Foryer G and Lefrant S 1999 *Macromolecules* **32** 6807
[11] Hotta S, Soga M and Sonoda N 1989 *J. Phys. Chem.* **93** 4994
[12] Gustafsson G, Inganas O, Osterholm H, Laakso J 1991 *Polymer* **32** 1574

Ambipolar Carrier Mobility in Binary Blend Thin Film of Non-Peripheral Alkylphthalocyanines

Akihiko Fujii[1], Shohei Nakano[1], Hitoshi Fukui[1], Takashi Saito[1], Masashi Ohmori[1], Yo Shimizu[2], and Masanori Ozaki[1]

[1]Division of Electrical, Electronic and Information Engineering, Graduate School of Engineering, Osaka University, Suita, Osaka 565-0871, Japan
[2]Synthetic Nano-Function Materials Group, Research Institute for Ubiquitous Energy Devices, Kansai Center, National Institute of Advanced Industrial Science and Technology (AIST), Ikeda, Osaka 563-8577, Japan

E-mail: afujii@eei.eng.osaka-u.ac.jp

Abstract. Charge carrier mobility in blend films of two types of soluble phthalocyanine derivatives with different substituent length, 1,4,8,11,15,18,22,25-octapentyl-phthalocyanine ($C5PcH_2$) and 1,4,8,11,15,18,22,25-octahexyl-phthalocyanine ($C6PcH_2$) has been investigated. The charge carrier mobility was measured by the time-of-flight technique. In the case of $C5PcH_2$ blend ratio below 25 mol%, the high mobility, such as 0.8-1.1 $cm^2V^{-1}s^{-1}$ for hole and 0.6 $cm^2V^{-1}s^{-1}$ for electron, were obtained at room temperature. In the thin films with $C5PcH_2$ above 30 mol%, the charge carrier mobility decreased by one order of magnitude and had the different temperature dependence from that below 25 mol%. The marked change of the charge carrier mobility depending on the blend ratio of phthalocyanine derivatives has been discussed by taking the miscibility and the molecular packing structure into consideration.

1. Introduction

Organic electronics is one of the most attractive and important research fields to realize light-weight, flexible and ubiquitous electronic devices in commercial base. In particular, a recent trend of this research field is "printed electronics" based on a solution processing, and realization of the low-cost organic thin film devices, such as organic light-emitting diodes, organic field effect transistors and organic photovoltaics, is promising. Required characteristics of organic semiconductors available for the printed electronics are high solubility in common organic solvents, spontaneous molecular alignment during the solvent evaporation to form a high quality thin film on a substrate, and high carrier mobility, and so on. Some organic semiconductor materials, such as pentacene derivatives [1,2], benzothienobenzothiophene derivatives [3] and dinaphthothienothiophene derivatives [4], were reported and must be the candidates applicable to the printed electronics.

Recently, we reported unique properties of a phthalocyanine derivateive, 1,4,8,11,15,18,22,25-octahexylphthalocyanine ($C6PcH_2$) [5]. $C6PcH_2$ is soluble in common organic solvents and has the discotic liquid crystalline (LC) properties. $C6PcH_2$ forms a hexagonal columnar structure by self-organization in the LC and crystal phases, and demonstrates the high ambipolar carrier mobility comparable to that of amorphous silicon, 1.08 and 0.43 $cm^2V^{-1}s^{-1}$ for hole and electron at room temperature, respectively. $C6PcH_2$ is available for bulk-heterojunction (BHJ) solar cells as a donor material and the high power conversion efficiency (PCE) of 3.2 % [6,7] were reported. In the C6PcH$_2$ based BHJ solar cells, the PCE could be improved to be 3.8% [8] using the blend donor materials of C6PcH$_2$ and its analogue molecule, 1,4,8,11,15,18,22,25- octapentylphthalocyanine (C5PcH$_2$).

Utilizing the mixture of small molecules with different alkyl-substituents, improvement of charge carrier mobility, and molecular alignment control have been reported [9-11]. The details of fundamental properties of the blend thin film of C5PcH$_2$ and C6PcH$_2$ should include an important essence to realize high performance devices, however, have not been clarified yet.

In this study, the charge carrier mobility in the blend films of C5PcH$_2$ and C6PcH$_2$ was investigated by a time-of-flight technique and discussed by taking the miscibility and molecular packing structure in the films.

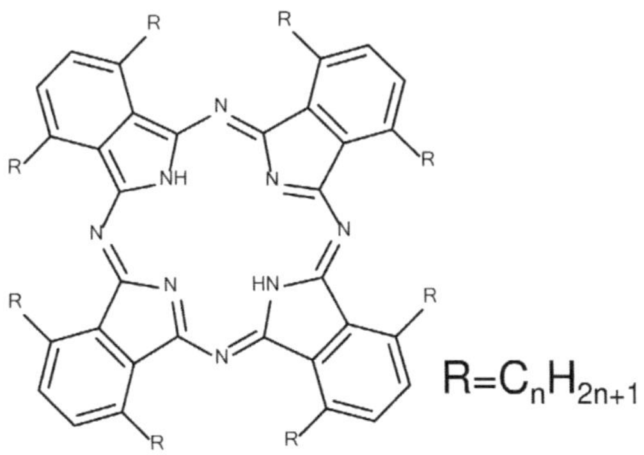

Figure 1. Molecular structures of C5PcH$_2$ (n=5) and C6PcH$_2$ (n=6).

2. Experimental

The molecular structures of C5PcH$_2$ and C6PcH$_2$ are shown in Fig. 1. C5PcH$_2$ and C6PcH$_2$ were synthesized according to the literature [12] with slight modifications, and fully purified by column chromatography (silica-gel with toluene as eluent) followed by repetitive recrystallization from toluene-methanol (1:2) solution.

The mixtures of C5PcH$_2$ and C6PcH$_2$ were prepared by the following procedures. C5PcH$_2$ and C6PcH$_2$ were dissolved in toluene, and toluene was vaporized under a N_2 atmosphere. The molar ratios of C5PcH$_2$ and C6PcH$_2$ were 12.5:87.5, 25:75, 30:70, 37.5:62.5, 50:50 and 75:25.

A sandwich cell, consisting of two indium tin oxide (ITO)-coated glass substrates and a 9-μm-thick PET spacer, was used for the TOF measurements. The actual cell gap was evaluated by an interference technique of light transmittance. The mixture of C5PcH$_2$ and C6PcH$_2$ was heated to a temperature in the isotropic phase for injecting into the cell by a capillary action, and cooled down to room temperature in a vacuum.

The cell was set onto a sample stage with a heater, which could control the sample temperature during the measurement by TOF technique, in a vacuum chamber. A DC bias voltage was applied between the electrodes using batteries connected in series and the sample was irradiated with a Nd:YAG pulsed laser (λ: 355 nm, pulsed width: 1 ns). The generated transient current was detected by an oscilloscope (LeCroy HDO 4054), which allowed the transit time to be determined, and the mobility μ was estimated according to the equation, $\mu = l^2/V\tau$, where l, V, and τ are the sample thickness, applied voltage and transit time, respectively.

The phase transition behavior of the mixtures of C5PcH$_2$and C6PcH$_2$ was investigated by differential scanning calorimetry (DSC, TA instruments Q2000) and polarizing optical microscopy (POM, Nikon ECLIPSE E600 POL) equipped with a Mettler FP82HT Hot Stage and Mettler FP90 Central Processor.

X-ray diffraction (XRD) pattern was measured using an X-ray diffractometer (Rigaku RINT 2000) with Cu Kα line. A measured sample was prepared by separating the pair of glass substrates of the sandwich cell. The lattice parameter d was calculated from the XRD patterns using Bragg's law; $n\lambda = 2d\sin\theta$, where n is an integer, λ is the X-ray wavelength of 1.5418 Å, and θ is the Bragg angle.

3. Results and Discussion

In order to confirm the miscibility of $C5PcH_2$ and $C6PcH_2$, the thermal properties in the mixture of $C5PcH_2$ and $C6PcH_2$ were investigated. Figure 2 shows the phase transition diagram determined by using the DSC and POM of the mixture of $C5PcH_2$ and $C6PcH_2$. A LC phase appears between 161 °C and 169 °C for $C6PcH_2$, while no LC phase exists in the case of $C5PcH_2$. The clearing points of the mixtures were kept constant in the case of $C5PcH_2$ blend ratio lower than 50 mol%. The melting points almost coincided with Le Chatelier-Schröder relation [13-15] exhibited as a theoretical thermal behavior of the melting point for each blend ratio. The eutectic point of the mixture could be approximately determined to be 20 mol% of the $C5PcH_2$ blend ratio from the calculation using Le Chatelier-Schröder relation. Therefore, these results indicate that the mixtures of $C5PcH_2$ and $C6PcH_2$ possess the miscibility, and each compound is mixed in the molecular level.

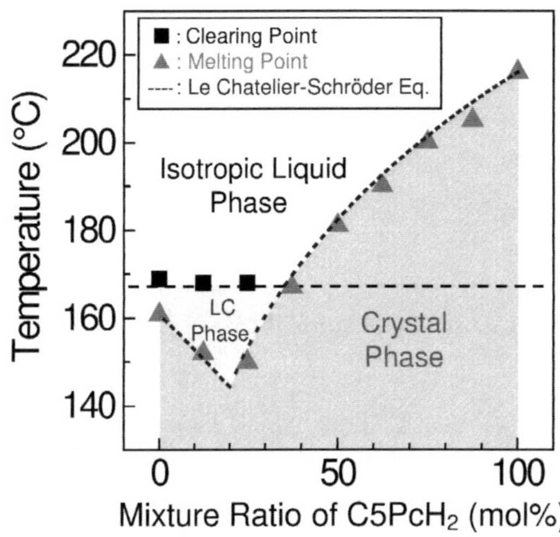

Figure 2. Phase transition diagram for the mixtures of $C5PcH_2$:$C6PcH_2$.

Figure 3 shows typical decay curves of the transient photocurrents for hole and electron in the blend film of $C5PcH_2$:$C6PcH_2$=50:50 under an applied electric field of 27.9 kV/cm at room temperature. The decay curves with the clear kink points were observed throughout the temperature range of the crystal phase and the transit times for hole and electron transport could be determined. The decay curves of other blend ratios of $C5PcH_2$ and $C6PcH_2$ were also nondispersive behaviors except for those of the electron in the blend ratios of 30:70 and 75:25. The charge carrier mobility was independent of the applied field, which is a typical behavior observed in liquid crystalline semiconductors [16].

Figures 4(a) and 4(b) show the temperature dependence of the hole and electron mobility in the blend films of $C5PcH_2$:$C6PcH_2$, respectively. Comparing the temperature dependence, the obvious different behaviors could be found among the three regions, that is, in the $C5PcH_2$ blend ratio below 25 mol%, 30-37.5 mol%, and 50-75 mol%. In the region of $C5PcH_2$ blend ratio below 25 mol%, the charge carrier mobility was independent of the temperature and approximately 0.25 $cm^2V^{-1}s^{-1}$ for hole and 0.3 $cm^2V^{-1}s^{-1}$ above about 140 °C, while below about 140 °C the charge carrier mobility increased from 0.5 to 1.0 $cm^2V^{-1}s^{-1}$ for hole and from 0.3 to 0.6 $cm^2V^{-1}s^{-1}$ for electron as the temperature decreased. In the region of 30-37.5 mol%, the charge carrier mobility slightly increased from 0.01 to 0.05 $cm^2V^{-1}s^{-1}$ with decreasing the temperature. In the region of 50-75 mol%, the charge carrier mobility was independent of the temperature and approximately 0.25 $cm^2V^{-1}s^{-1}$ for hole and 0.3 $cm^2V^{-1}s^{-1}$ above about 140 °C, while below about 140 °C the charge carrier mobility increased from 0.05 to

0.1 $cm^2V^{-1}s^{-1}$ with decreasing the temperature. The temperature dependence of the charge carrier mobility in crystal phase could be explained by the volume expansion of the unit lattice in the molecular packing structure. The detailed studies about the temperature dependence of the lattice parameter in the $C5PcH_2$:$C6PcH_2$ blend single crystal by using X-ray structure analysis are now in progress, and the relationship between the mobility and lattice parameter would be reported elsewhere.

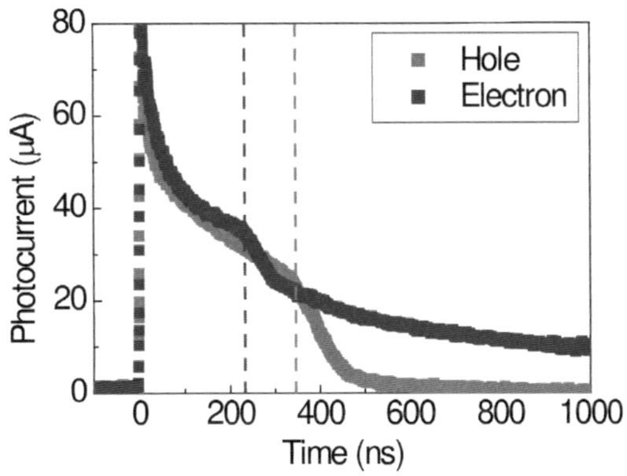

Figure 3. Transient photocurrent decay curves of the blend film of $C5PcH_2$:$C6PcH_2$=50:50 under the field of 27.9 kV/cm at room temperature.

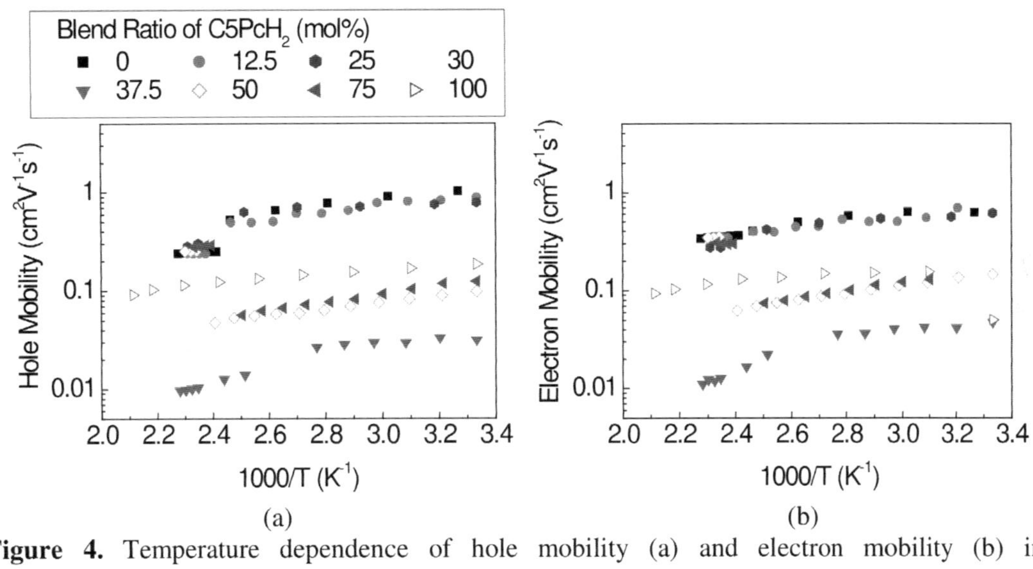

Figure 4. Temperature dependence of hole mobility (a) and electron mobility (b) in the $C5PcH_2$:$C6PcH_2$ blend films.

Comparing the charge carrier mobility in the blend films of $C5PcH_2$:$C6PcH_2$, the $C5PcH_2$ blend ratio dependence of the mobility at room temperature is shown in Fig. 5. In the case of $C5PcH_2$ blend ratio below 25 mol%, both hole and electron mobilities kept high values close to those of $C6PcH_2$. The hole mobility slightly decreased from 1.1 to 0.8 $cm^2V^{-1}s^{-1}$ with increasing the blend ratio of $C5PcH_2$, while the electron mobility kept constant, 0.6 $cm^2V^{-1}s^{-1}$. In the case of $C5PcH_2$ blend ratio above 30 mol%, both hole and electron mobilities markedly decreased by one order of magnitude, and the hole mobility slightly increased with increasing the blend ratio of $C5PcH_2$.

Figures 6(a) and 6(b) show the XRD patterns of the blend films with $C5PcH_2$:$C6PcH_2$ and the magnified figure in the range of 2θ=4-6°, respectively. In the case of $C5PcH_2$ blend ratio below 25

Journal of Physics: Conference Series **704** (2016) 012006 doi:10.1088/1742-6596/704/1/012006

mol%, the diffraction peaks at 4.86° and 26.02° appeared, which correspond to the *d*(100) lattice parameter of hexagonal column formed by C6PcH_2 (Fig. 7(a)) and the *d*(001) lattice parameter of inter-molecular distance, respectively. It is considered that the molecular packing structure like C6PcH_2 was formed and the molecules were stacked in order, since C6PcH_2 was dominant for the self-organization in the appearance of the liquid crystal phase in the C5PcH_2 blend ratio below 25 mol%, therefore, the charge carrier mobility was kept high.

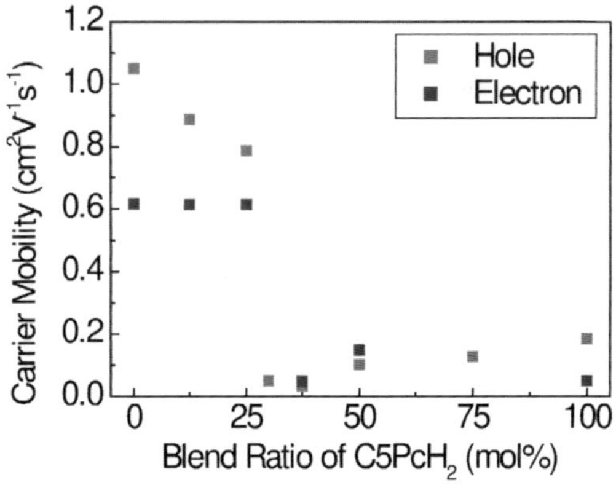

Figure 5. C5PcH_2 blend ratio dependence of charge carrier mobility in the C5PcH_2:C6PcH_2 blend films at room temperature.

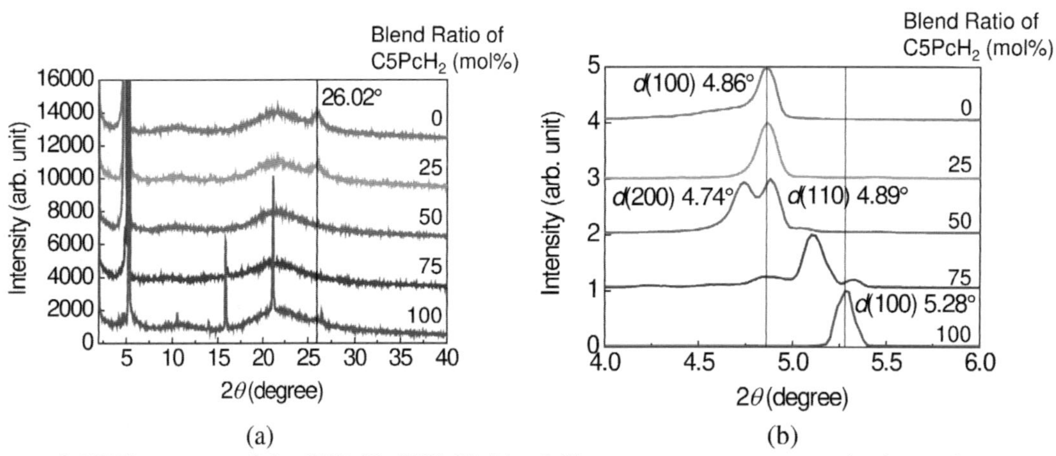

Figure 6. XRD patterns of the C5PcH_2:C6PcH_2 blend films at room temperature in the angle ranges of 2-40° (a) and 4-6° (b).

In the C5PcH_2 blend ratio of 50 mol%, the two diffraction peaks at 4.74° and 4.89° were observed, which correspond to the *d*(200) and *d*(110) lattice parameter of rectangular column (Fig. 7 (b)). It is supposed that the lattice constant in the *a*-axis direction of rectangular lattice became longer, while that in the *b*-axis direction became shorter, because equivalent mixing caused the distortion of the hexagonal lattice.

In the case of C5PcH_2 blend ratio above 50 mol%, the diffraction peak shifted from 4.89° to 5.28° corresponding to the *d*(100) in Fig. 7(a) as the blend ratio of C5PcH_2 increased. It is considered that the mean length of the alkyl chains around the column became shorter with increasing the blend ratio of C5PcH_2, because of short substituent of C5PcH_2, resulting in the decrease of the inter-columnar distance.

The diffraction peaks around 26°, which originates from the molecular stacking, however, disappeared in the higher C5PcH$_2$ blend ratio than 25 mol%, therefore, the marked decrease of the carrier mobility in the film with C5PcH$_2$ above 30 mol% must be explained by the disordered stacking in the columnar structure. Since the inter-molecular overlapping of π-orbitals must strongly contribute to the carrier transport, therefore, it is considered that the carrier transport in the blend film of C5PcH$_2$ and C6PcH$_2$ should be dominated by the order of molecular packing like π-stacking rather than the lattice parameter like inter-columnar distance.

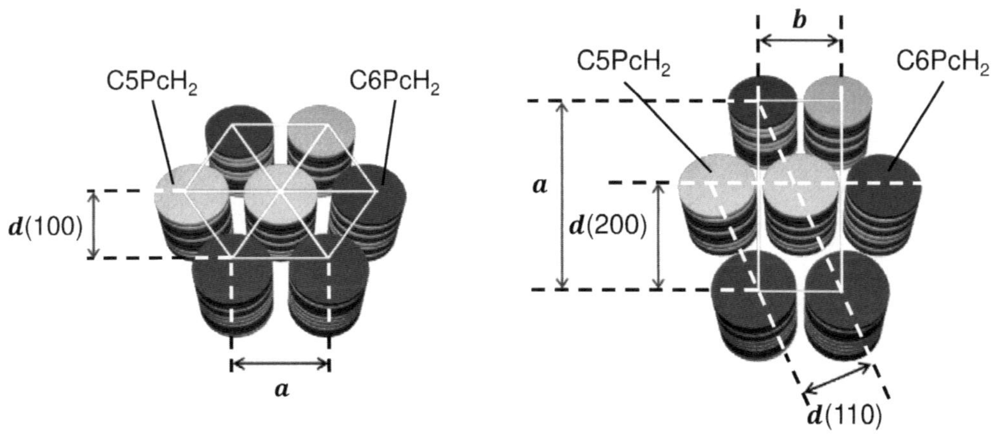

(a) (b)

Figure 7. Schematic diagrams of (a) hexagonal and (b) rectangular columnar structures in the blend films of C5PcH$_2$:C6PcH$_2$.

4. Conclusions

The charge carrier mobility in the blend films of C5PcH$_2$ and C6PcH$_2$ were investigated. The phase diagram of the mixture of C5PcH$_2$ and C6PcH$_2$ was determined from the thermal properties, it was found that the mixture possess the miscibility. In the case of C5PcH$_2$ blend ratio below 25 mol%, both hole and electron mobilities were kept high values comparable with those of C6PcH$_2$, while both hole and electron mobilities markedly decreased by one order of magnitude in the case of C5PcH$_2$ blend ratio above 30 mol%. From the results of XRD measurement, the molecular structure of the mixture film obviously changed depending on the blend ratio of C5PcH$_2$, that must be strongly related with the behavior of the charge carrier transport.

Acknowledgements

This work was partly supported by JSPS KAKENHI Grant Numbers 15H03552, 26600073, 24246009, and by Advanced Low Carbon Technology Research and Development Program from the Japan Science and Technology Agency (JST-ALCA), and by Photonics Advanced Research Center Program in Osaka University.

References

[1] Chang J, Sakanoue T, Olivier Y, Uemura T, Dufourg-Madec M, Yeates S G, Cornil J, Takeya J, Troisi A and Sirringhaus H 2011 *Phys. Rev. Lett.* **107** 066601
[2] Llorenete G R, Dufourg-Madec M, Crouch D J, Printchard R G, Ogier S and Yeates S G 2009 *Chem. Commun.* 3059
[3] Uemura T, Hirose Y, Uno M, Takimiya K and Takeya J 2009 *Appl. Phys. Express* **2** 111501

[4] Nakayama K, Hirose Y, Soeda J, Yoshizumi M, Uemura T, Uno M, Li W, Kang M J, Yamagishi M, Okada Y, Miyazaki E, Nakazawa Y, Nakao A, Takimiya K and Takeya J 2011 *Adv. Mater.* **23** 1626
[5] Miyake Y, Shiraiwa Y, Okada K, Monobe H, Hori T, Yamasaki N, Yoshida H, Cook M J, Fujii A, Ozaki M and Shimizu Y 2011 *Appl. Phys. Express* **4** 021604
[6] Hori T, Miyake Y, Yamasaki N, Yoshida H, Fujii A, Shimizu Y and Ozaki M 2010 *Appl. Phys. Express* **3** 101602
[7] Hori T, Fukuoka N, Masuda T, Miyake Y, Yoshida H, Fujii A, Shimizu Y and Ozaki M 2011 *Sol. Energy Mater. Sol. Cells* **95** 3087
[8] Fukui H, Nakano S, Uno T, Dao Q, Saito T, Fujii A, Shimizu Y and Ozaki M 2014 *Org. Electron.* **15** 1189
[9] Wegewijs B R and Siebbeles L D A 2002 Phys. Rev. B **65** 245112
[10] Shimizu Y, Matsuda Y, Nekelson F, Miyake Y, Yoshida H, Fujii A and Ozaki M 2009 *Proc. of SPIE* **8279** 82790G
[11] G. Schweicher, G. Gbabode, F. Quist, O. Debever, N. Dumont, S. Sergeyev and Y. H. Geerts: 2009 *Chem. Mater.* **21** 5867
[12] Swarts J C, Langner E H G, Krokeide-Hove N and Cook M J 2001 *J. Mater. Chem.* **11** 434
[13] Le Chatelier H 1894 *C. R. Acad. Sci.* **118** 638
[14] Schröder I 1893 *Z. Phys. Chem.* **11** 449
[15] Van Laar J J 1903 *Arch. Neerl.* **8** 264
[16] Yoshino K, Nakayama H, Ozaki M, Onoda M and Hamaguchi M 1997 *Jpn. J. Appl. Phys.* **36** 5183

Electrophoretically deposited multiwalled carbon nanotube based amperometric genosensor for *E.coli* detection

Hema Bhardwaj, Shipra Solanki and Gajjala Sumana*

Biomedical Instrumentation Section, CSIR-National Physical Laboratory, New Delhi-110012, India.

E-mail: sumanagajjala@gmail.com

Abstract. This work reports on a sensitive and selective genosensor fabrication method for Escherichia coli (*E.coli)* detection. The functionalized multiwalled carbon nanotubes (MWCNT) synthesized via chemical vapour deposition have been deposited electrophoretically onto indium tin oxide coated glass surface and have been utilized as matrices for the covalent immobilization of *E. coli* specific probe oligonucleotide that was identified from the 16s rRNA coding region of the *E. coli* genome. This fabricated functionalized MWCNT based platform sought to provide improved fundamental characteristics to electrode interface in terms of electro-active surface area and diffusion coefficient. Electrochemical cyclic voltammetry revealed that this genosensor exhibits a linear response to complementary DNA in the concentration range of 10^{-7} to 10^{-12} M with a detection limit of 1×10^{-12} M.

1. Introduction

The detection of bacterial pathogens in food and water is of extreme importance as food borne and water borne diseases have become a major public health problem worldwide due to increasing pollution and changing lifestyle [1]. The presence of *Escherichia coli* (*E. coli*) is used as a potential marker for diagnoses of water borne diseases affecting humans [7]. Most of the strains of *E. coli* are harmless but some strains are pathogenic and cause diarrhoea, urinary tract infection, inflammation and peritonitis. The traditional methods for *E. coli* detection include multiple-tube fermentation, membrane filter and plate counts that are accurate and sensitive, but these techniques have certain limitations such as operational complications, lack of specificity and are time consuming (24–48 hrs) [8]. Therefore, rapid and sensitive technique for *E. coli* detection is urgently needed.

Electrochemical DNA biosensors based on the conversion of molecular recognition reactions into useful response signals, offer considerable promise for DNA analysis. Different nanomaterials have been utilized for matrix formation of various biosensors. The biomedical applications of nanoscale materials are governed by the fact that they exhibit size compatibility with bio-receptors such as cells, viruses, proteins and nucleic acids. There has been an explosion of interest in the use of carbon based nanomaterials for biosensor development owing to their properties like large surface area, conductivity, biocompatibility and ease of functionalization. Different carbon materials which have been studied for biosensing application include graphite powder, carbon nanotubes (CNTs), fullerenes, carbon nanofibers and graphene. Of these

CNTs are of special interest as they possess all the properties like large surface area, high conductivity, biocompatibility and their special geometry allows suitable biomolecule immobilization [4]. Two types of carbon nanotubes are known, single walled (SWCNTs) and multiwalled (MWCNTs). SWCNTs possess a singular graphite sheet rolled into a tube to create a cylindrical nanostructure, whereas MWCNTs consist of several shells of cylindrical tubes. Both MWCNTs and SWCNTs have been used to fabricate sensitive and selective enzyme based biosensors, immunosensors as well as genosensors [11, 12].

For biosensing application, a precise control over the spatial arrangement and distribution of the nanoparticles onto the substrate is utmost important. Among the various techniques, electrophoretic deposition (EPD) offers rapid deposition of desired materials and their composites with a simple design setup, thereby presenting a scope as an automated and continuous process on an industrial scale [2]. It provides tunable control over the degree of deposition of the charged colloidal particles onto an electrode surface under the influence of an applied electric field. In the present work, we have utilized electrophoretically deposited MWCNTs platform for the fabrication of genosensor for *E. coli* detection. In order to increase the biomolecules immobilization as well as dispersability of MWCNTs to render it suitable for EPD they have been functionalized with –COOH groups. Then a specific probe oligonucleotide that is identified from the 16s rRNA coding region of the *E. coli* genome has been immobilized covalently onto the fabricated matrix to construct a label free electrochemical genosensor for sensitive and selective detection of *E.coli.*

2. Materials and methods

2.1 Chemicals

Toluene, Ferrocene, Bovine Serum Albumin (BSA), N-hydroxysuccinimide (NHS), 1-Ethyl-3-(3-dimethylaminopropyl) carbodiimide (EDC), potassium ferrocyanide ($K_4[Fe(CN)_6]$), potassium ferricyanide ($K_3[Fe(CN)_6]$), sodium dihydrogen phosphate (NaH_2PO_4), sodium monohydrogen phosphate (Na_2HPO_4) and all other chemicals were of analytical grade and were procured from Sigma Aldrich (www.sigmaaldrich.com). Aqueous solutions and buffers were prepared in Milli-Q water (18 MΩ cm). Probe sequence specific to *E. coli* that identified from the 16s rRNA coding region of the *E. coli* genome and complementary target sequences were procured from Sigma Aldrich, Milwaukee, USA and are as follows:13 DNA probe (pDNA): NH_2-5′-GGT CCG CTT GCT CTC GC-3′ complementary (cDNA): 5′-GCG AGA GCA AGC GGA CC-3′. The solutions of oligonucleotides were prepared in Tris–EDTA buffer (1 M Tris–HCl, 0.5 M EDTA) at pH 8.0 and stored at -20 °C prior to use.

2.2 Synthesis and Functionalization of carbon nanotubes

Multiwalled Carbon nanotubes have been synthesized using chemical vapour deposition (CVD). Toluene and ferrocene have been used as precursors that act as catalyst and hydrocarbon source respectively [6]. Further, synthesized carbon nanotubes have been functionalized by using a refluxing process. In a 100 mL of round bottom flask, MWCNTs have been dispersed in 25 ml of concentrated nitic acid by continuous magnetic stirring for 12 hours at 80 °C, resulting in generation of -COOH group on the MWCNT surface [3]. Finally, functionalized MWCNTs have been washed with distilled water to remove the impurities from the solution by maintaining upto neutral pH and the sample is dried at room temperature.

2.3 Electrophoretic Deposition

Firstly, indium tin oxide (ITO) coated glass substrates have been hydrolysed by using $H_2O_2/NH_4OH/H_2O$ solution in the ratio 1:1:5 v/v for 45 min at 80 °C. The solution of CNTs has been prepared in distilled water then sonicated for 2 hrs. For electrophoretic deposition technique, a constant DC voltage source for electrophoretic deposition having two electrodes is used. A surface charge on MWCNT has been generated by mixing a magnesium nitrate solution into CNTs solution which acts as an electrolyte. ITO coated glass electrode and platinum electrode act as cathode and anode respectively by separation 1 cm is immersed into the suspension solution of CNTs. Electrophoretically films have been generated onto ITO electrode by applying 40V for 60 seconds. The fabricated MWCNT/ITO electrode was then removed from the suspension and washed with distilled water followed by the drying at room temperature. The procedure for EPD is shown in Figure 1(A).

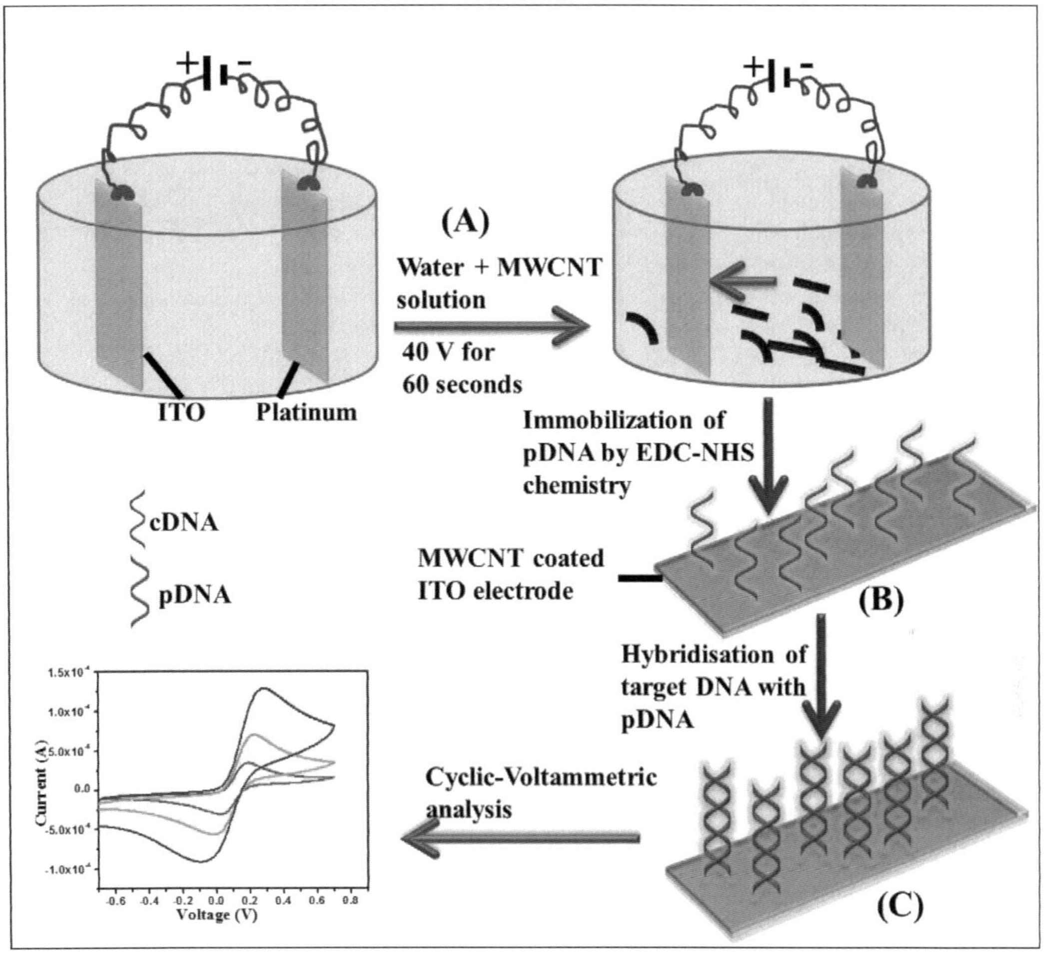

Figure 1. Schematic showing (A) electrophoretic deposition of MWCNTs onto ITO coated glass substrate, (B) covalent immobilization of pDNA onto the MWCNTs electrode, and (C) hybridisation of target DNA with pDNA on bioelectrode.

2.4 Immobilization of biomolecules onto fabricated electrode surface

The fabricated MWCNT/ITO electrode have been activated using EDC (2 mM) and NHS (5 mM) and kept for 1 hr in the dark. Subsequently, 40 ml of pDNA (10^{-7} M) have been immobilized onto the modified electrode at 100% humidity at room temperature (~27 °C) for about 6 hrs, followed by rinsing with Tris–HCl. The hybridization studies have been performed as a function of target DNA concentration with complementary sequences in a humid chamber for about 20 min at 35 °C. The proposed mechanism for the synthesis of MWCNT/ITO and the fabrication of the nucleic acid sensor for *E. coli* detection is shown in Figure 1(B and C).

3. Characterization

Spectroscopic and morphological characterization has been carried out for the structural analysis using UV-Visible spectrophotometer (Lambda 950, Perkin Elmer), Fourier transform infrared (FT-IR) spectroscopy (Spectrum BX, Perkin- Elmer) and transmission electron microscopy (HR-TEM, Tecnaii-G2F30 STWIN). Contact angle (CA) measurements have been obtained using a contact angle meter (Data Physics OCA15EC). Zeta potential measurements have been done using Malvern zetasizer instrument (Zetasizer Nano, ZS90). Electrochemical characterization has been carried out by using Autolab Potentiostat/Galvanostat (Eco Chemie, The Netherlands) in phosphate buffer saline (PBS, 50 Mm, 0.9% NaCl, pH -7.4) containing 5 mM $(Fe(CN)6)^{3-/4-}$ using three electrode cell having Ag/AgCl as reference electrode and platinum as the counter electrode.

4. Results and discussion

4.1 *Spectroscopic and morphological analysis of material synthesized*

UV-visible spectra of MWCNTs have recorded and shown in Figure 2(A). The absorbance peak of the absorbance spectra was observed at 280 nm. The absorbance peaks observed at 280 nm is mainly due to π-π^* electronic transition of C-C bonds in MWCNTs by dispersion of sample in distilled water.

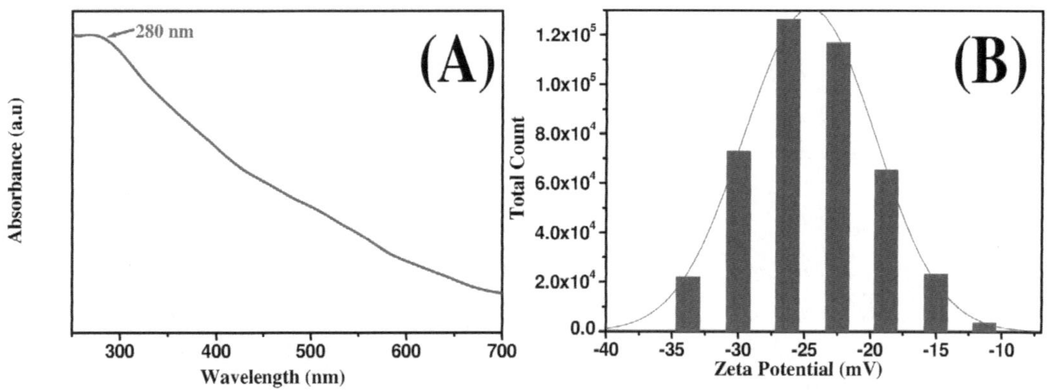

Figure 2. (A) UV-visible spectra of MWCNT and (B) Zeta potential analysis of the MWCNT colloid dispersed in water.

Figure 3 shows the Fourier transform infrared spectra (FT-IR) of (A) MWCNT/ITO and (B) pDNA/MWCNT/ITO electrodes. The FT-IR spectrum of MWCNT/ITO exhibits broad peak at 3000 cm^{-1} corresponding to -OH symmetric stretching in carboxylic acids showing MWCNTs have been functionalized by -COOH groups. The other peak has been observed at 1342 cm^{-1} mainly due to the C=C, which shows that synthesized MWCNTs have been functionalized by carboxylic group. Further, pDNA have been immobilized onto MWCNT/ITO electrode surface which shows characteristic peak at 1600

cm^{-1} which corresponds to N-H stretch of amide-I as shown in curve (Fig. 3B). The additional sharp peak observed at 3600 cm^{-1} which corresponds to N-H stretch of amide bond (1°amine) indicating the formation of amide linkage by immobilization of pDNA onto carboxylated MWCNT/ITO coated electrode surface which confirms the covalent attachment of DNA with MWCNTs surface.

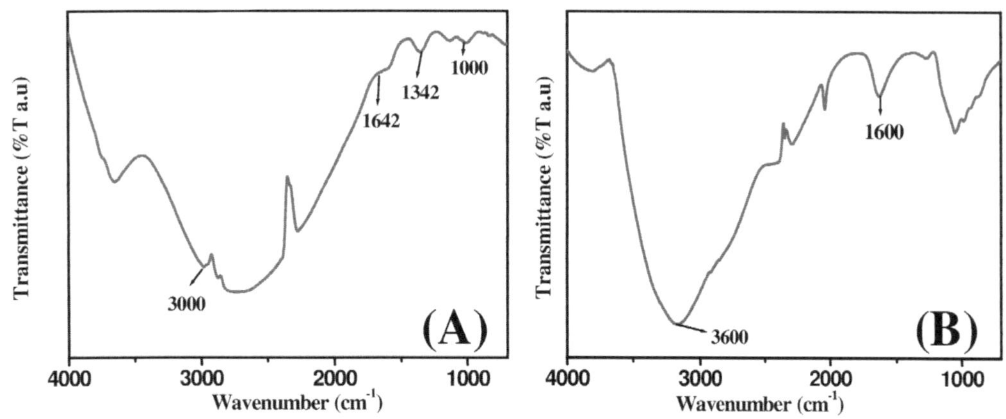

Figure 3. FT-IR spectra of (A) MWCNT /ITO electrode and (B) pDNA/MWCNT/ITO bioelectrode.

Figure 4 shows high resolution transmission electron microscopy (HR-TEM) studies has been carried for the structural investigation of synthesized MWCNTs. The uniformly distributed long tube like structures have been obtained having uniform diameter without any splintering in a long nanotube. The Fig. 4(A) shows distinct formation of long tubes of multiwalled CNTs which easily distinguish by the TEM images that confirmed the formation of multiwalled carbon nanotube. The Fig. 4(B) shows that multiwalled carbon nanotube have been formed and marked by red line which shows multiwalled formation take place. The Fig. 4(C) shows the lattice fringes of MWCNTs with an interplanar spacing 1.0A° indicating high crystallinity in carboxylated carbon nanotubes. The diameter of MWCNTs varies from 50 to 80 nm with length.

4.2 Zeta potential analysis

To determine the particle stability of functionalized MWCNTs zeta potential measurements have been conducted at 25°C. All the experiments have been repeated three times and standard deviation has been calculated. It was observed that MWCNTs dispersed in Milli Q water at pH 7 after sonication for 30 minutes show an average zeta potential of -24.7 mV (Fig. 2 (B)) having a standard deviation of 0.23 indicating that the colloid solution formed is quite stable and suitable for EPD.

4.3 Contact angle analysis of electrodes

To investigate the film formation and DNA immobilization onto the ITO surface, contact angle measurements have been carried out using sessile drop method (Fig. 5). The decrease in value from 59.87° for the MWCNT/ITO electrode (Fig. 4B) to 35.37° for pDNA/MWCNT/ITO (Fig. 5C) electrode indicates the increase in the number of hydrophilic functional groups (-COOH, NH_2 and –SH) on the surface and thereby confirming the DNA immobilization.

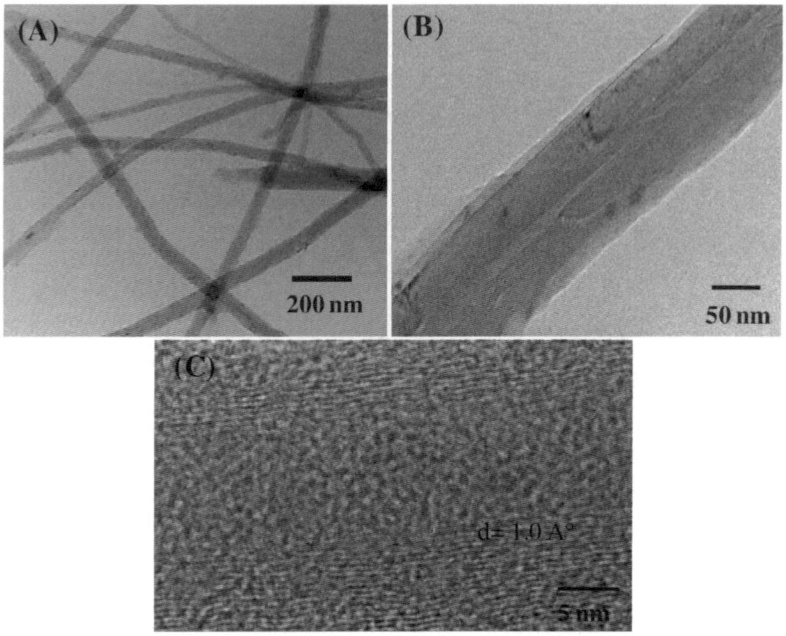

Figure 4. TEM images of (A) Carbon nanotubes, (B) High resolution TEM image of Single carbon nanotube showing different walls, and (C) High resolution image showing crystalline fringes of the synthesized multiwalled carbon nanotubes.

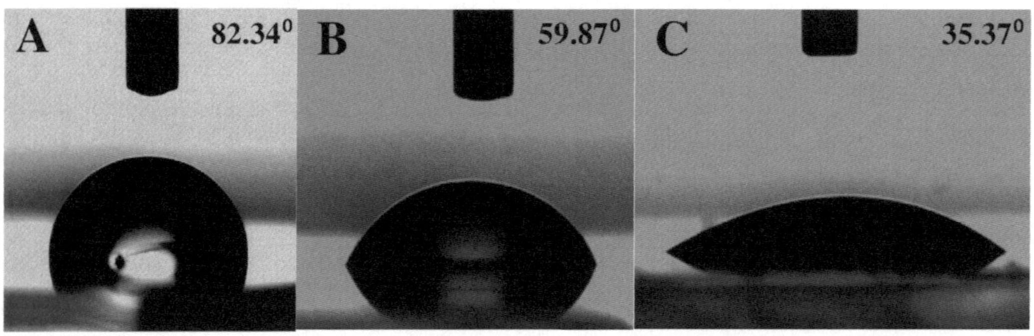

Figure 5. Contact angle images of (A) ITO coated glass surface (B) MWCNT coated ITO electrode (C) pDNA immobilized electrode.

4.4 Electrochemical characterization of electrodes
The electrochemical studies of different electrodes have been carried out by dipping the electrodes in 13 mL of phosphate buffer saline (PBS) (100 mM, pH 7.4, 0.9% NaCl) containing 5 mM $[Fe(CN)_6]^{3-/4-}$ using cyclic voltammetry (CV). CV analysis of bare ITO electrode, MWCNT/ITO electrode, and pDNA/MWCNT/ITO bioelectrode has been carried out at scan rate (υ) of 50 mV/s. Figure 6(B) shows anodic peaks at 0.3078 mA, 0.1435 mA and 0.0704 mA for ITO, MWCNT/ITO and

pDNA/MWCNT/ITO electrodes, respectively. The decrease of anodic peak current of MWCNT/ITO electrode as compared to ITO indicates successful formation of film and further decrease in the current after pDNA immobilization is due to the hindrance in the diffusion of electrons caused due to immobilization of large biomolecule.

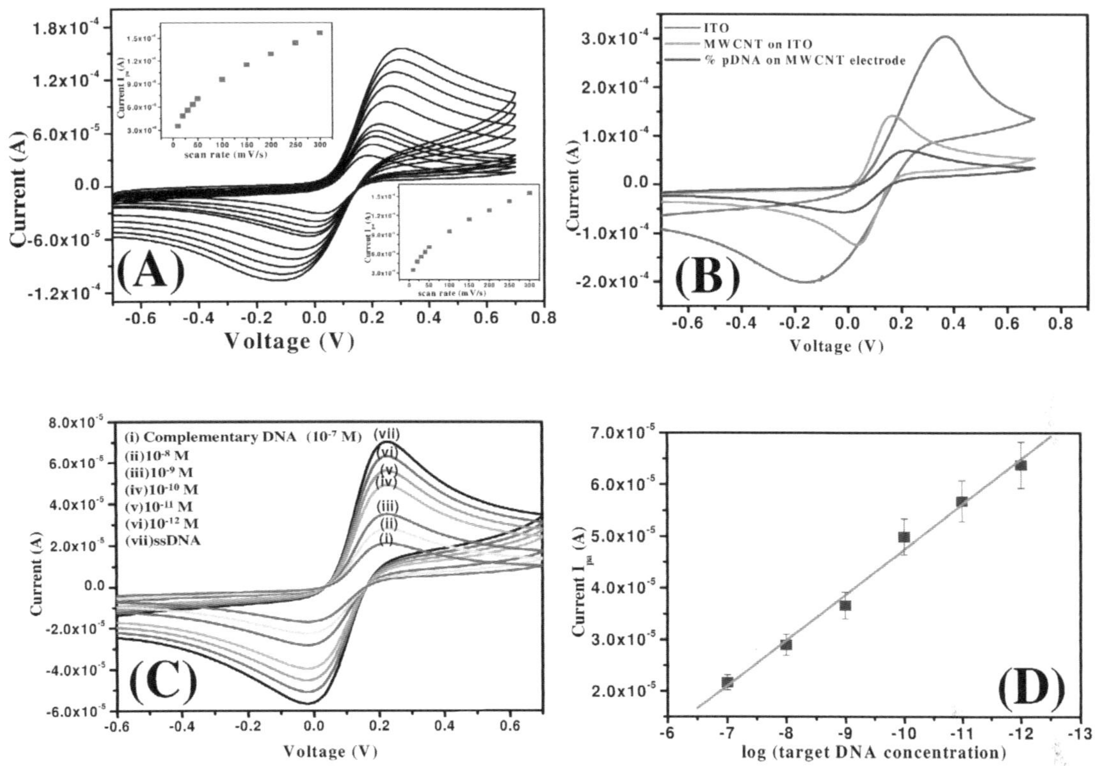

Figure 6. (A) Cyclic voltammetry response studies of pDNA/MWCNT/ITO bioelectrode at different scan rates (10-300 mV/s) in 5 mM $[Fe(CN)_6]^{3-/4-}$ PBS solution at pH 7 (inset (i) variation of anodic peak current I_{pa} (A) with scan rate υ (mV/s) and (ii) variation of anodic peak potential E_{pa} with log of scan rate υ (mV/s), (B) CV studies of different substrates, (C) Biosensing studies of the pDNA/MWCNT/ITO bioelectrodes at different cDNA concentrations, (D) Variation of anodic peak current I_{pa} with cDNA concentration peak current (I_{pa}) concentration.

The CV studies of pDNA/MWCNT/ITO bioelectrode at different scan rates υ (10–300 mV/s) indicate that as we move towards higher scan rate, anodic peak current (I_{pa}) shifts towards more positive value and cathodic peak current (I_{pa}) shifts in the reverse direction respectively. Besides this, the redox peak currents show linear behaviour as a function of $_{\text{scan rate}}$ ($\upsilon^{1/2}$) (Fig. 6A inset (i) and (ii)) revealing it as a diffusion controlled electron-transfer process and is given by the following equations.

I_{pa} (A) (pDNA/MWCNT/ITO) = 0.0045 + 0.4009 (μAs/mV) × scan rate (mV/s);
R = 0.9823 (1)

E_{pa} (V) (pDNA/MWCNT/ITO) = 0.1008 + 0.0764 (s) × log of scan rate (mV/s);
R= 0.9883 (2)

The diffusion coefficient of the pDNA/MWCNT/ITO bioelectrode has been determined using the Randle Sevick equation:

$$I_{pa} = (2.69\times10^{5})n^{3/2} AD^{1/2} C \upsilon^{1/2} \qquad (3)$$

Where, I_{pa} is the peak current (I_{pa} anodic and I_{pc} cathodic), n is the number of electrons, A is the area of electrode (0.5 cm^2), D is the diffusion coefficient, C is the surface concentration, and v is the scan rate (50 mV/s). The D value has been obtained as 1.24×10^{-7} cm^2/s. The total surface concentration of probe DNA is calculated using Laviron's theory and is found to be 1.65×10^{-10} mol cm^{-2}, indicating high surface coverage of DNA onto MWCNT/ITO electrode [5]. The activity of the pDNA/MWCNT/ITO bioelectrode has been estimated as a function of pH varying from 6.0 to 8.0 at 25 °C. The high magnitude of response current obtained at pH 7.0 (data not shown) indicates that pDNA/MWCNT/ITO bioelectrode is more active at pH 7.0, at which DNA molecules retain their natural structure and do not get denatured [9]. Thus, all experiments have been conducted at pH 7.0 at 25 °C.

4.5 Biosensing studies

Figure 6(C) represents the CV studies of pDNA/MWCNT/ITO bioelectrode carried out at different concentrations of complementary DNA (10^{-12} M to 10^{-7} M) by immersing the electrode in PBS containing 5 mM $[Fe(CN)_6]^{3-/4-}$. It has been observed that there is linear decrease in the anodic peak current with increase in analyte concentration indicating that decrease in the charge transfer to the working electrode. This mainly due to the increase in the concentration of cDNA results in more hybridization of heavy biomolecules with the immobilized probe take place and the hindrance in the charge transfer to the electrode increases. This causes decrease in the resulting current. Figure 6(D) shows variation in anodic The equation for line of regression for linear region is given by:

I_{pa}(A) = 0.0165+ 8.778 (μA) × log of conc. of cDNA (M);
R=0.9987 (4)

The limit of detection (LOD) calculated for linear region using the expression 3σ/sensitivity [10], where σ is the standard deviation (n = 10) of the CV signals obtained in the absence of cDNA is 1×10^{-12} M.

5. Conclusions

In summary, we demonstrate the studies of electrochemical genosensor for *E.coli* detection based on the functionalized multiwalled carbon nanotubes. Uniformly formed functionalized multiwalled CNTs have been synthesized using chemical vapour deposition technique and deposited onto indium tin oxide (ITO) coated glass substrates using electrophoretic deposition (EPD) method. The fabricated MWCNT/ITO electrode surfaces have been utilized for the immobilization of pDNA of *E. coli* through covalent bonding. The detection using an electrochemical cyclic voltammetry analysis reveals that the fabricated biosensor with an appropriately optimized protocol can accurately detect *E. coli* in the range of 10^{-7} to 10^{-12} M. Our sensing strategy could be extended for the detection of other micro-organisms and open up new avenues for the designing of electrochemical nucleic acid sensors.

Acknowledgements

We thank Director, CSIR-NPL, New Delhi, India for providing the facilities to work. Shipra Solanki is thankful to UGC, India, for the award of JRF. Hema Bhardwaj is thankful to CSIR-NPL, New Delhi, India for providing financial support. We thank Mr. Dinesh Singh (CSIR-NPL, India) for TEM studies.

We acknowledge Dr. A.M.Biradar (CSIR-National Physical Laboratory), Dr. R.K.Kotnala (CSIR-National Physical Laboratory) and Prof. Bansidhar Malhotra (Delhi Technological University) for useful discussions. Authors are thankful to Mr. M.K.Pandey for suggestions and support. Mr. Chandan Singh and all members of Biomedical Instrumentation Section, CSIR-NPL, New Delhi are acknowledged for interesting discussions. The financial support received from DST and CSIR, India is highly acknowledged.

References

[1] Cabral J o P S 2010 Water microbiology. Bacterial pathogens and water *Int. J. Environ. Res. Public Health* 7 3657-703

[2] Corni I, Ryan M P and Boccaccini A R 2008 Electrophoretic deposition: from traditional ceramics to nanotechnology *J. Eur. Ceram. Soc.* 28 1353-67

[3] Datsyuk V, Kalyva M, Papagelis K, Parthenios J, Tasis D, Siokou A, Kallitsis I and Galiotis C 2008 Chemical oxidation of multiwalled carbon nanotubes *Carbon* 46 833-40

[4] Hu C and Hu S 2009 Carbon nanotube-based electrochemical sensors: principles and applications in biomedical systems *J. Sens.* 2009

[5] Laviron E, Roullier L and Degrand C 1980 A multilayer model for the study of space distributed redox modified electrodes: Part II. Theory and application of linear potential sweep voltammetry for a simple reaction *J. Electroanal. Chem. Interfacial Electrochem.* 112 11-23

[6] Mathur R B, Chatterjee S and Singh B P 2008 Growth of carbon nanotubes on carbon fibre substrates to produce hybrid/phenolic composites with improved mechanical properties *Composites Science and Technology* 68 1608-15

[7] Pandey C M, Sharma A, Sumana G, Tiwari I and Malhotra B D 2013 Cationic poly (lactic-coglycolic acid) iron oxide microspheres for nucleic acid detection *Nanoscale* 5 3800-7

[8] Pandey C M, Sumana G and Tiwari I 2014 Nanostructuring of hierarchical 3D cystine flowers for high-performance electrochemical immunosensor *Bios. and Bioelectron.* 61 328-35

[9] Thomas P S 1980 Hybridization of denatured RNA and small DNA fragments transferred to nitrocellulose proceedings of the *Nat. Acad. Sci.* 77 5201-5

[10] Vidal J C, García-Ruiz E and Castillo J R 2001 Design of a multilayer cholesterol amperometric biosensor for preparation and use in flow systems *Electroanalysis* 13 229-35

[11] Wang J, Liu G and Jan M R 2004 Ultrasensitive electrical biosensing of proteins and DNA: carbon-nanotube derived amplification of the recognition and transduction events *J. Am. Chem. Soc.* 126 3010-1

[12] Zhu N, Chang Z, He P and Fang Y 2005 Electrochemical DNA biosensors based on platinum nanoparticles combined carbon nanotubes *Anal. chim. acta* 545 21-6

Polaron states and charge dynamics in doped conjugated polymers studied by various spectroscopic methods

Hideo Kishida

Department of Applied Physics, Nagoya University, Furo-cho, Chikusa-ku, Nagoya 464-8603, Japan

E-mail: kishida@nuap.nagoya-u.ac.jp

Abstract. Various spectroscopic data for doped conjugated polymers are reviewed and discussed. Chemical doping induces polaron states in the conjugated polymers. Two polaron states appear between the π and π* states. Third harmonic generation spectroscopy reveals the multiply-resonant nonlinear optical processes using the polaron states. The terahertz measurements clarify the carrier localization, which is dependent on the doping levels. The ultrafast luminescence measurements reveal the relaxation process of the photoexcited states of the doped conjugated polymers.

1. Introduction

In organic electronics and nanotechnology, including the fields of organic conductors, electronic luminescent devices, field effect transistors (FET), and photovoltaic devices, mobile charges play important roles. The mobile charges often appear in the partially occupied π states of organic electronic systems. For example, in organic FET, the gate voltage controls the charge filling in the highest occupied molecular orbital (HOMO) or the lowest occupied molecular orbital (LUMO) [1] and realizes the partial occupation of these orbitals. In doped conjugated polymers, chemical doping gives the two polaron states (P1 and P2) in which the partially occupied π states are achieved [2].

Polarons induced by doping affect and change the pattern of the carbon skeleton in the conjugated chains, which leads to the emergence of new electronic states within the π-π* gap. The spatial extent and anisotropy of the polarons are crucial to the usage of the polaron carriers. Moreover, it is important to understand the charge dynamics in the states where the polarons are on the conjugated chains. In this paper, the author reviews the electronic states in doped polythiophene, which is one of the most popular conjugated polymers, revealed by various optical experiments using femtosecond lasers. In the third-harmonic generation (THG) spectroscopy, the transition dipole moments between the polaron states and π (π*) states are evaluated [3]. In the terahertz (THz) spectroscopy, the optical conductivity in the THz region is discussed [4–6]. Moreover, in the time-resolved luminescence measurement, the initial dynamics of the photoexcited states of the doped conjugated polymers are discussed [7]. From these optical properties of the doped states of the conjugated polymers, the nature of the carriers is discussed.

2. Doped poly(3-hexylthiophenes) [3, 4, 6]

The doped states of poly(3-hexylthiophenes) (P3HT), which are one of the main targets in this paper, are introduced in this section. Similar to most undoped conjugated polymers, the undoped P3HT has a

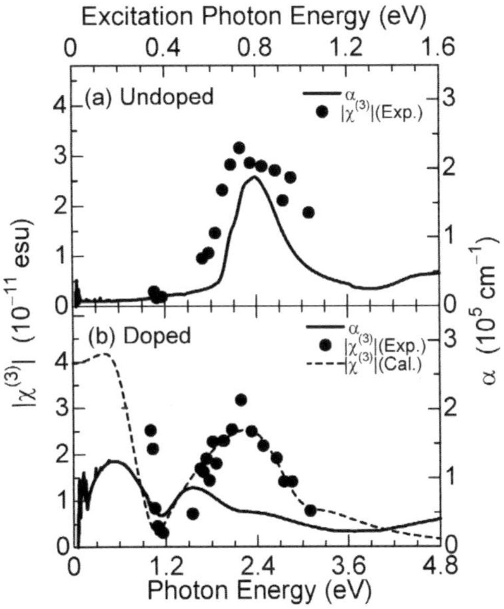

Figure 1. Absorption and THG spectra of (a) undoped and (b) doped P3HT. Reprinted with permission from H. Kishida *et al.*, Appl. Phys. Lett. **97**, 103302 (2010). Copyright 2010 American Institute of Physics.

strong absorption peak, which is assigned to the π-π* transition and is located at around 2.4 eV, as shown in figure 1(a) [3]. Chemical doping of P3HT is done by dipping the polymer into an acetonitrile solution of $Cu(ClO_4)_2$ both for regioregular and regiorandom P3HT. Even after taking the sample out of the dopant solution, the samples will remain stable for a sufficient amount of time for optical measurements. By changing the immersion time, we can control the degree of doping. The following laser experiments are performed for the samples which are taken out of the solution and placed in the air. By the chemical doping of P3HT, the absorption spectra related to the polaron states appear below the π-π^* transition and are shown in figure 1(b). π-P1 transitions are observed at around 0.6 eV and the P1-P2 transitions are at around 1.5 eV.

3. Spectroscopy for the polaron states

3.1. Third harmonic generation measurements [3]

In the THG experiments, a pulsed laser is irradiated on the sample and the generated third harmonic (TH) light is detected. The intensity of the TH light is compared with that of the reference sample and the third-order nonlinear susceptibility $|\chi^{(3)}|$ is obtained. The spectra of $|\chi^{(3)}|$ for the undoped and doped P3HT are shown in figure 1 [3]. The upper axis is the photon energy of the excitation laser, while the lower axis is the photon energy of the TH light. The absorption spectra are also plotted against the same lower axis. In the undoped sample, the peak positions of $|\chi^{(3)}|$ and absorption in figure 1(a) are the same, indicating that three-photon resonance occurs. For the doped sample, the peak energy positions of $|\chi^{(3)}|$ and that of the absorption are different. As the result of spectral analysis, one-photon resonance

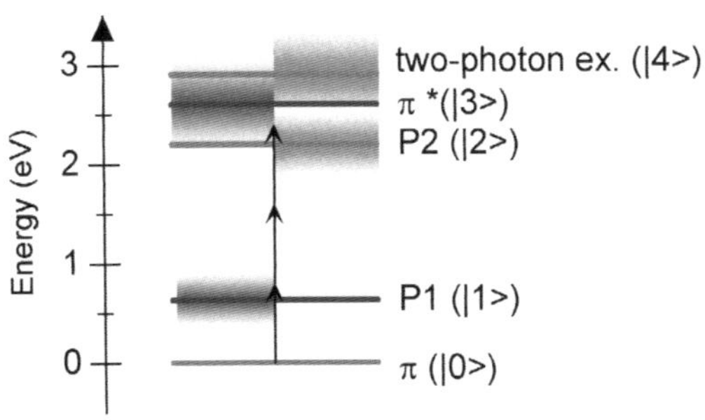

Figure 2. Energy level structure of doped P3HT. Reprinted with permission from H. Kishida *et al.*, Appl. Phys. Lett. **97**, 103302 (2010). Copyright 2010 American Institute of Physics.

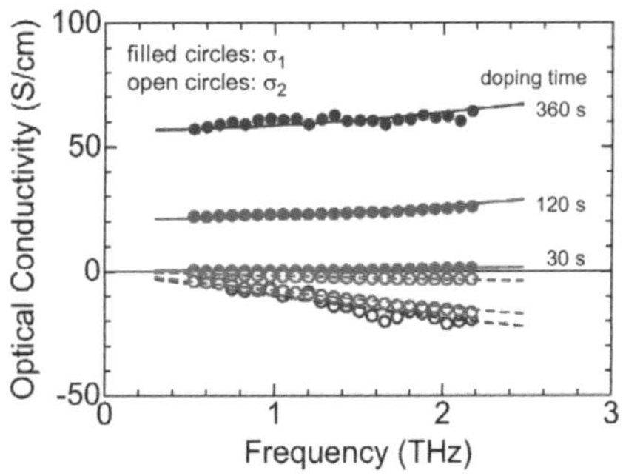

Figure 3. THz spectra of doped P3HT. Reprinted with permission from T. Unuma *et al.*, Appl. Phys. Lett. **97**, 033308 (2010). Copyright 2010 American Institute of Physics.

and three-photon resonance doubly occur as shown in figure 2 [3] and the spectral shape is determined. The spectral analysis also reveals the transition dipole moments between π, π^*, and the two polaron states. These experimental data reveal that the polaron states can contribute to the third-order nonlinear optical processes.

3.2. THz spectroscopy [4–6]

The dynamics of the charges in the doped conjugated polymers should be reflected in the low-frequency optical spectra. The optical conductivity spectra of the conjugated polymers, which are obtained from the time-domain THz spectroscopy, are discussed in this subsection. The THz spectra for the doped P3HT are shown in figure 3 [4]. With increased doping time, the absolute values of the optical conductivities also increase. The spectra are analyzed by the Drude-Smith model [8], in which the backward scattering of the carriers are considered. Such scattering leads to the localization of the carriers. In the analysis, the localization parameter C ($-1 \leq C \leq 0$), carrier density N, and the relaxation time τ are calculated. $C = -1$ gives complete localization, while $C = 0$ is complete delocalization, where the model coincides with the normal Drude model. The data for regiorandom P3HT and PEDOT:PSS samples are also compared in figure 4. At low carrier density, the localization parameter is close to -1, which indicates that the states are insulating. Even at the highest carrier density, C is below -0.5, which is in the region of localized states.

The anisotropy of the optical conductivity was reported for the oriented P3HT films fabricated by the rubbing method [6]. The optical conductivity parallel to the rubbing direction is significantly larger than that perpendicular to the rubbing direction. This fact indicates that the rubbing procedure aligns the polymer chains and that the charge motion governing the conductivity in the THz regions is along the polymer chain.

The THz measurements were extended to highly conducting polyanilines (PA) [5]. A highly conducting PA shows large optical conductivity values of over 500 S/cm. As a result of the spectral analysis by the Drude-Smith model, C reaches -0.23. This value indicates that the localization mechanism is significantly weakened and that the true metallic state is realized in the highly conducting PA.

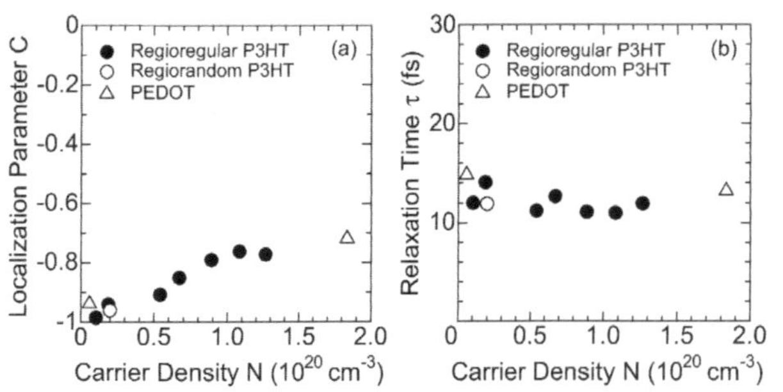

Figure 4. (a) Localization parameter C and (b) relaxation time τ versus carrier density. Reprinted with permission from T. Unuma *et al.*, Appl. Phys. Lett. **97**, 033308 (2010). Copyright 2010 American Institute of Physics.

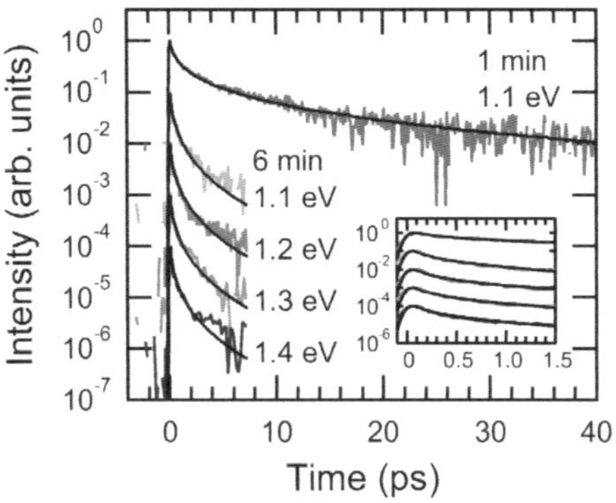

Figure 5. The time evolution of the luminescence intensity of doped P3HT measured by the up-conversion method. Reprinted in part with permission from T. Koyama, ACS Photon. **1**, 655 (2014). Copyright 2014 American Chemical Society.

3.3. Luminescence and up-conversion measurements [7, 12]
Non-doped conjugated polymers such as polythiophenes are often luminescent [9]. On the other hand, the luminescence of the doped state is significantly quenched [10]. The quenching mechanism is not clear at present, but it is easily expected that the polaron states can be a path for the energy relaxation of the photoexcited states and one of the quenchers of the luminescence. However, luminescence from the doped states of P3HT was observed [7]. From the energy of the luminescence, it is assigned to the P2-P1 transition. The decay of the luminescence for the doped P3HT was measured using the up-conversion method. In the experiment, the excitation was done with an 80 fs laser pulse at 1.55 eV and the luminescence was probed at 1.1 to 1.4 eV. The sum frequency light of the luminescence and the gating pulse (1.55 eV, 80 fs) was detected. The time evolution of the luminescence (figure 5 [7]) is reproduced by a stretched exponential function. The decay time evaluated in the analysis of stretched exponential functions is approximately 30 fs. From the analysis, the mobility of the charges is estimated to be 2 $cm^2V^{-1}s^{-1}$. This value is one order of magnitude larger than the macroscopically measured value [11]. The large mobility observed in the luminescence measurement reflects the fact that the up-conversion measurement detects the initial motion of the photocarriers and that various trapping or scattering processes do not work in such an ultrafast time region.

As for the PEDOT:PSS, a similar time evolution was observed [7]. Therefore, the relaxation process is inherent to the polythiophene backbone. In addition to the polaron luminescence, PEDOT:PSS shows the luminescence in the visible region under 3.08 eV excitation, which is assigned to the luminescence from the dopant polymer PSS [12]. This visible luminescence should be considered when we fabricate visible optical devices using PEDOT:PSS.

4. Conclusions
The polaron states and the charge dynamics in the doped conjugated polymers are discussed on the basis of the experimental results of various spectroscopic methods. The THG spectroscopy reveals the multiple resonance using both π (π^*) states and polaron states. This clarifies that the polaron states and π (π^*) states are not spatially separated, but that the π (π^*) states and polaron states are significantly overlapped. The THz spectroscopy demonstrates the localization of the carriers in doped P3HT and the true metallic state in highly conducting polyaniline. These results suggest that the THz method is useful for the judgement of the metallic nature of the carriers. The time-resolved luminescence measurement for doped states clarifies the dynamics of the photoexcited states of the doped conjugated polymers. Thus, the various optical measurements using femtosecond lasers are a powerful tool for unravelling the physics of polarons in doped conjugated polymers.

Acknowledgements
The author thanks T. Unuma, T. Koyama, Y. Yokoyama, A. Nakamura, K. Hibino, K. Fujii, N. Yamada, A. Umemoto, T. Matsuno, S.-C. Lee, E.-Y. Hong, S.-H. Lee, and O-P. Kwon for their collaboration. He is also grateful for the support from KAKENHI (17067005 and 22340079) and Ogasawara Foundation for the Promotion of Science & Engineering.

References
[1] Newman C R, Frisbie C D, da Silva Filho D A, Brédas J L, Ewbank P C and Mann K R 2004 *Chem. Mater.* **16** 4436
[2] Brédas J L and Street G B 1985 *Acc. Chem. Res.* **18** 309
[3] Kishida H, Hibino K and Nakamura A 2010 *Appl. Phys. Lett.* **97** 103302
[4] Unuma T, Fujii K, Kishida H and Nakamura A 2010 *Appl. Phys. Lett.* **97** 033308
[5] Unuma T, Yamada N, Nakamura A, Kishida H, Lee S C, Hong E Y, Lee S H and Kwon O P 2013 *Appl. Phys. Lett.* **103** 053303
[6] Unuma T, Umemoto A and Kishida H 2013 *Appl. Phys. Lett.* **103** 213305
[7] Koyama T, Nakamura A and Kishida H 2014 *ACS Photon.* **1** 655
[8] Smith N V 2001 *Phys. Rev.* B **64** 155106
[9] Perepichka I F, Perepichka D F, Meng H and Wudl F 2005 *Adv. Mater.* **17** 2281
[10] Arkhipov V I and Bässler H 2004 *Phys. Status Solidi* a **201** 1152
[11] Jiang Y D, Jen T H and Chen S A 2012 *Appl. Phys. Lett.* **100** 023304
[12] Koyama T, Matsuno T, Yokoyama Y and Kishida H 2015 *J. Mater. Chem.* C **3** 8307

Journal of Physics: Conference Series **704** (2016) 012009 doi:10.1088/1742-6596/704/1/012009

Amphiphilic Fluorinated Polymer Nanoparticle Film Formation and Dissolved Oxygen Sensing Application

Yu Gao, Huie Zhu, Shunsuke Yamamoto, Tokuji Miyashita, Masaya Mitsuishi

Institute of Multidisciplinary Research for Advanced Materials, Tohoku University, 2-1-1 Katahira, Aoba-ku, Sendai 980-8577, Japan

E-mail: masaya@tagen.tohoku.ac.jp

Abstract. Fluorinated polymer nanoparticle films were prepared by dissolving amphiphilic fluorinated polymer, poly(*N*-1*H*,1*H*-pentadecafluorooctylmethacrylamide) (pC7F15MAA) in two miscible solvents (AK-225 and acetic acid). A superhydrophobic and porous film was obtained by dropcasting the solution on substrates. With higher ratios of AK-225 to acetic acid, pC7F15MAA was densified around acetic acid droplets, leading to the formation of pC7F15MAA nanoparticles. The condition of the nanoparticle film preparation was investigated by varying the mixing ratio or total concentration. A highly sensitive dissolved oxygen sensor system was successfully prepared utilizing a smart surface of superhydrophobic and porous pC7F15MAA nanoparticle film. The sensitivity showed I_0/I_{40} = 126 in the range of dissolved oxygen concentration of 0 ~ 40 mg L^{-1}. The oxygen sensitivity was compared with that of previous reports.

1. Introduction

In the past decades, polymer nanoparticles have attracted much attention in polymer science.[1-3] The field of polymer nanoparticles is expanding and playing an important role in fundamental and practical research areas, ranging from sensors, electronics, photonics, conducting materials, medicine, biotechnology, pollution control, to environmental technology. Among several types of polymer nanoparticles, prepolymer nanoparticles have been intensively investigated.

Recently we demonstrated porous nanoparticle film preparation using amphiphilic fluorinated polymers.[4] We introduced an enormously enhanced dissolved oxygen sensor system consisting of amphiphilic acrylamide-based polymers: poly(*N*-1*H*, 1*H*-pentadecafluorooctylmethacrylamide) (pC7F15MAA) and poly(*N*-dodecylacrylamide-*co*-5-[4-(2-methacryloyloxyethoxycarbonyl)phenyl]-10,15,20-triphenylporphinato platinum(II)) (p(DDA/PtTPP)) (Figure 1). The nanoparticle formation ability ensures both superhydrophobicity with a water contact angle larger than 160° and gas permeability so that molecular oxygen can enter the film from water. The film was prepared by casting a mixed solution of pC7F15MAA and p(DDA/PtTPP) with AK225 and acetic acid onto a solid substrate. The film has a porous structure made of nanoparticle assemblies with diameters of several hundreds of nanometers. The film shows exceptional performance as the oxygen sensitivity reaches 126 (the intensity ratio at two different oxygen concentrations (I_0/I_{40}); 40 and 0 (mg L^{-1}) correspond to dissolved oxygen concentration). Herein the nanoparticle film formation was investigated by varying the mixing ratio and concentration of the mixed solution. The dissolved oxygen sensitivity is also discussed by comparing the result with those of previous reports.

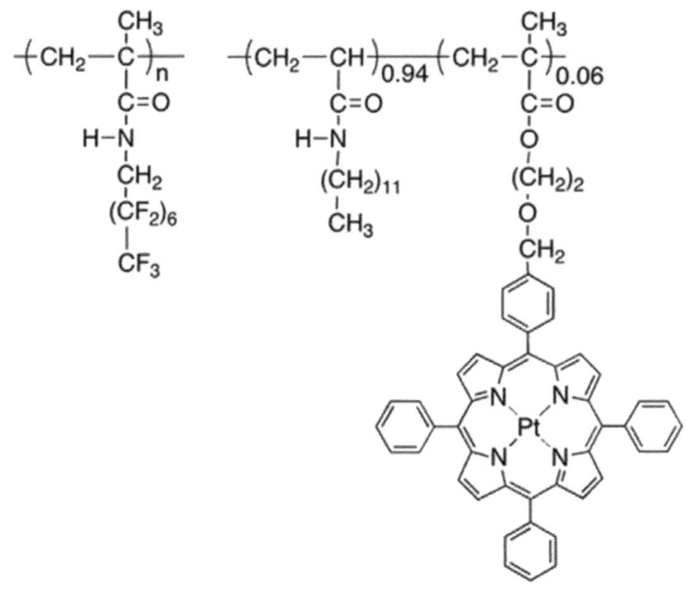

Figure 1. Chemical structures of pC7F15MAA and p(DDA/PtTPP).

2. Experimental Section

2.1. Materials

1*H*,1*H*-Pentadecafluorooctylamine was purchased from SynQuest Labs., Inc. Methacryloyl chloride was purchased from Wako Pure Chemical Industries, Ltd. AK-225 (a mixture of $CF_3CF_2CHCl_2$ and $CClF_2CF_2CHClF$) was purchased from Asahi Glass Co., Ltd. Hexane, dodecane, and tetradecane were purchased from Kanto Chemical Co., Inc. *n*-Octane was purchased from Tokyo Chemical Industry Co., ltd. Decane was purchased from Wako Pure Chemical Industries, Ltd. All chemicals were used without purification.

Amphiphilic fluorinated polymer of poly(*N*-1*H*, 1*H*-pentadecafluorooctylmethacrylamide) (pC7F15MAA, Figure 1) was prepared by free radical polymerization in AK-225. The polymer was reprecipitated in the excess amount of chloroform three times. Amphiphilic Pt-labeled porphyrin copolymer (p(DDA/PtTPP)) was prepared by free radical polymerization in toluene as described in a previous report.[5] The amount of PtTPP was determined by 6 mol% using UV spectroscopy, assuming that the mole extinction coefficient of PtTPP at 402 nm is 3.08×10^5 $M^{-1}cm^{-1}$.[6]

2.2. Preparation of pC7F15MAA nanoparticle film

For preparing a porous film consisting of pC7F15MAA nanoparticle assemblies, we used two types of solvents; AK-225 and acetic acid. AK-225 serves as a good solvent while acetic acid serves as a poor solvent. A mixed solution of pC7F15MAA and p(DDA/PtTPP) (95:5 w/w) in AK-225 and acetic acid was casted on solid substrates such as silicon wafer and glass substrate at room temperature. The cast film was kept at room temperature for at least one day to remove solvents.

2.3. Luminescence measurement

Luminescence spectra were measured using a fluorospectrophotometer (F-4500, Hitachi Ltd.). The sensing system was put in a quartz cell filled with water. The luminescence spectra were measured by varying a mixing ratio of argon and oxygen through a portable gas mixture device (PMG-1, Kofloc). The dissolved oxygen concentration was controlled by bubbling the water at a certain oxygen concentration for 10 min. The dissolved oxygen concentration was monitored using a dissolved oxygen meter (SG6, Mettler-Toledo International Inc). The measurement was carried out at room temperature.

2.4. Lifetime measurement
Time-resolved luminescence lifetime measurements were done using a single photon counting system. The sample was put in an in-house temperature-controlled pressure chamber. A Q-switched Nd:YAG laser (5 ns pulse duration, 10 Hz and 355 nm) was used as an excitation source. Luminescence decay data were obtained using a streak camera system (C4334, Hamamatsu Photonics K.K.). The decay curves were fitted with single exponential function. For the lifetime measurements in water, we set the sample in a quartz cuvette filled with water. The measurements were conducted after bubbling the water at a certain oxygen concentration for at least 10 min.

3. Results and Discussion

3.1. Porous pC7F15MAA nanoparticle film preparation

3.1.1. Effect of the mixing ratio
We examined how the experimental condition affect the nanoparticle film formation. First a mixed solution of pC7F15MAA in acetic acid and AK-225 with different mixing ratios (6:1, 5:1, 4:1, 3:1, 2:1, 1:1, 1:3, 1:4, 1:5, 1:6 v/v) was casted on solid substrates such as silicon wafer and glass substrate at room temperature. The cast film was kept at room temperature for at least one day to remove solvents. SEM images show that porous nanoparticle films were observed for the mixing ratio of acetic acid : AK-225 = 1:3 (Figure 2). Nanoparticle assembly formation is uniform on the whole substrate surface. For the mixing ratio of 1:5, however, inhomogeneous nanoparticle film formation was observed (Figure 2(b)). Though nanoparticles were generated, they took no film formation; the substrate was exposed in part. By screening the mixing ratio, it was found that no porous nanoparticle film formation was observed for the mixing ratios 1:5, 1:6, and 6:1 (Table 1). AK-225 is miscible with acetic acid. In the case of the higher mixing ratio of acetic acid (6:1), the effect of poor solvent on the solubility of pC7F15MAA is too strong. The pC7F15MAA cannot be dissolved in the mixed solvent immediately as AK-225 becomes vaporized. Regarding the nanoparticle size, the size became larger for 1:5 and 1:6 mixing ratios. In such situation, pC7F15MAA chains take more expanded conformations and undergo entanglements more easily as the solution is concentrated because of the solvent vaporization. Consequently pC7F15MAA nanoparticles with larger diameters (> 1μm) were generated and they took aggregates, resulting in inhomogeneous surface coating.

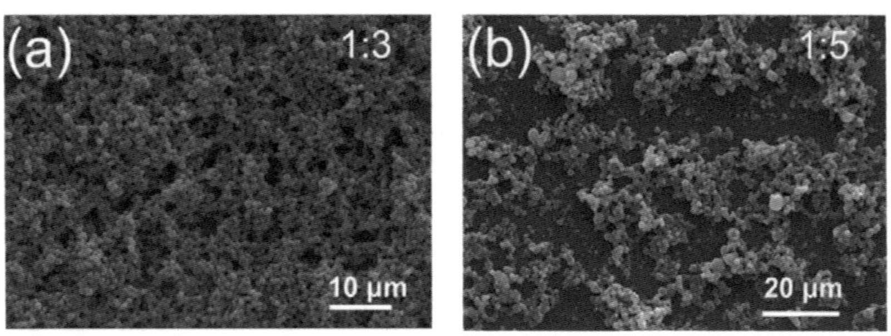

Figure 2. SEM images of pC7F15MAA nanoparticle films with a mixing ratio $V_{acetic\ acid} : V_{AK\text{-}225}$ = (a) 1:3 and (b) 1:5.

Table 1. Film formation ability as a function of the mixing ratio.

$V_{acetic\ acid} : V_{AK\text{-}225}$	6:1	5:1	4:1	3:1	1:1	1:2	1:3	1:4	1:5	1:6
Film quality	×	○	○	○	○	○	○	○	×	×

○ good; × poor

3.1.2. Effect of concentration
The effect of pC7F15MAA concentration was also examined with a constant volume ratio of 3:1. Figure 3 shows SEM images of pC7F15MAA nanoparticle films at concentrations of 0.25, 0.5, 0.75, and 1 wt%. Results show that a well-defined nanoparticle film was formed at a concentration of 0.5 wt% (Figure 3(b)). At lower concentration, pC7F15MAA nanoparticle films were covered by pC7F15MAA "sheet" (thin film) (Figure 3(a)). However, at high concentrations crack formation was observed in pC7F15MAA nanoparticle films (Figures 3(c) and 3(d)). Results imply that pC7F15MAA nanoparticle film formation is closely related to the solvent vaporization and that there is an optimized concentration: 0.5 wt%.

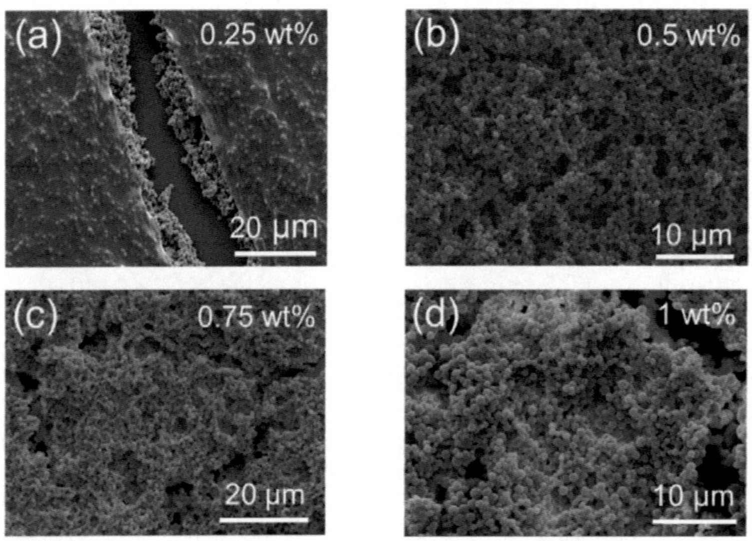

Figure 3. SEM images of pC7F15MAA nanoparticle films with different concentrations; (a) 0.25 wt%, (b) 0.5 wt%, (c) 0.75 wt%, and (d) 1 wt%.

3.1.3. Size distribution of pC7F15MAA nanoparticles
Figure 4 shows SEM images of pC7F15MAA nanoparticle films at the different mixing ratios of 5:1, 3:1, and 1:4. There is larger size distribution for 5:1 (Figure 4(a)). It becomes smaller as the AK-225 composition increases. For the mixing ratio 1:4 (Figure 4(c)), pC7F15MAA nanoparticles with the size ranging 400 - 700 nm were observed.

3.1.4. Surface wettability
The surface wettability pC7F15MAA nanoparticle films (acetic acid: AK-225 = 1:4) was examined using various solvents. Depending on the solvents, the films have different contact angles. Figure 5 shows contact angles of pC7F15MAA nanoparticle films. The water contact angle reaches more than 160°, indicating that the surface of pC7F15MAA nanoparticle films is superhydrophobic. Interestingly

Journal of Physics: Conference Series **704** (2016) 012009 doi:10.1088/1742-6596/704/1/012009

Figure 4. SEM images of pC7F15MAA nanoparticle films with different volume ratios: Acetic acid: AK-225 (v/v) = (a) 5:1, (b) 3:1, and (c) 1:4.

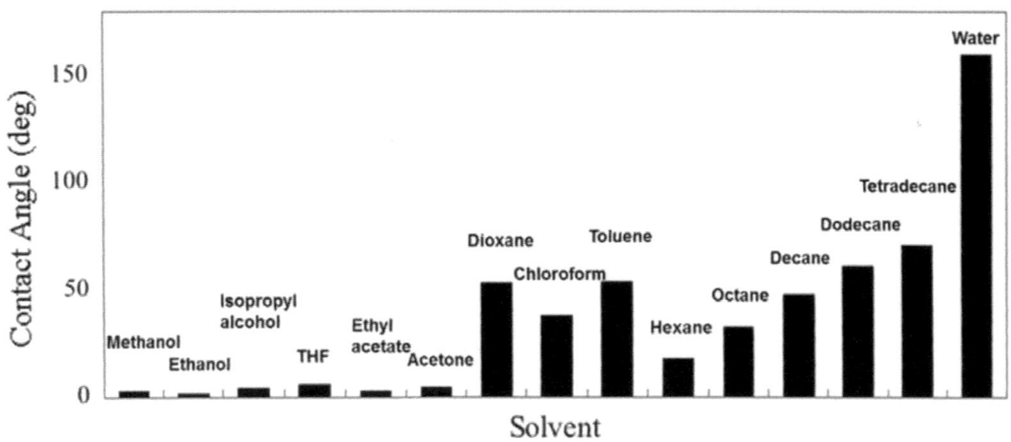

Figure 5. Surface wettability of pC7F15MAA nanoparticle films.

the surface of pC7F15MAA nanoparticle films shows quite low contact angle for methanol, ethanol, isopropyl alcohol, THF, ethyl acetate, and acetone. This indicates that pC7F15MAA nanoparticle film has high potential of solvent selective separation.

Next the critical surface free energy of pC7F15MAA nanoparticle film was investigated by measuring contact angles of five types of alkane solvents: hexane, decane, dodecane, octane and tetradecane (Figure 6). The critical surface free energy of pC7F15MAA nanoparticle film was determined as 18.6 mN m^{-1} (4:1) and 18.7 mN m^{-1} (1:5), respectively. This means that the surface of pC7F15MAA nanoparticle film was occupied by fluorocarbon groups.

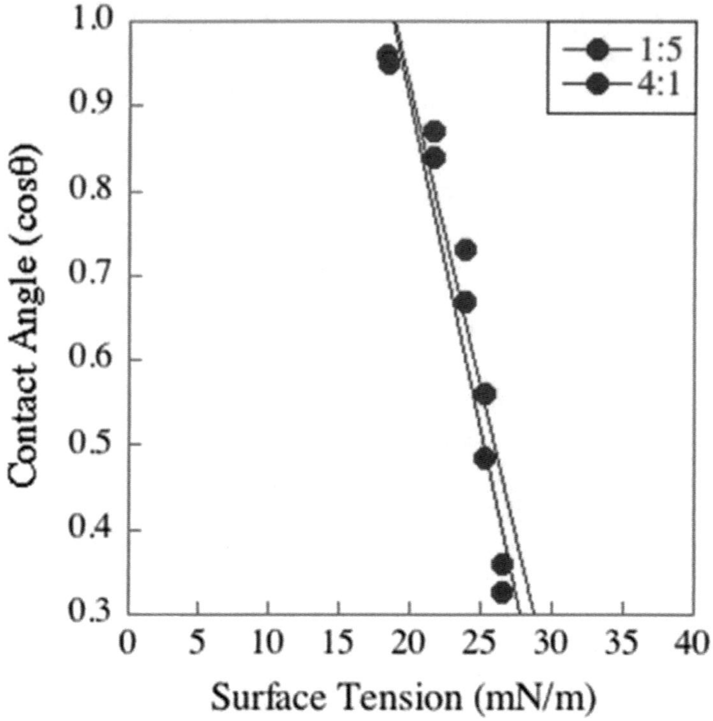

Figure 6. Zisman plot of pC7F15MAA nanoparticle film in the volume of 1:4 and 5:1.

3.2. Dissolved oxygen sensor performance

As described in Ref. [4], we have succeeded in preparing a high sensitive dissolved oxygen sensor system. Briefly, we prepared a porous and superhydrophobic pC7F15MAA nanoparticle film with platinum(II) porphyrin containing copolymer, poly(*N*-dodecylacrylamide-*co*-5-[4-(2-methacryloxyethoxycarbonyl)phenyl]-10,15,20-triphenylporphinatoplatinum(II)]) (p(DDA/PtTTP)). A

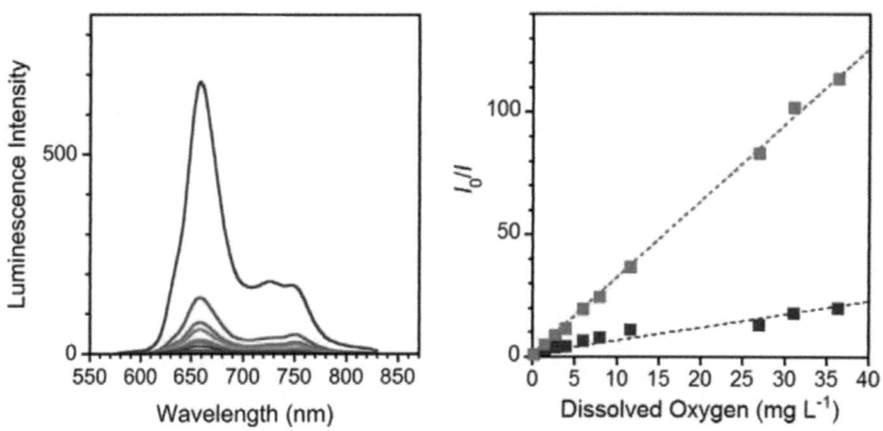

Figure 7. (left) Luminescence spectra and (right) Stern-Volmer plots of pC7F15MAA/p(DDA/PtTPP) nanoparticle film in water. Reprinted in part with permission from Ref. [4]. Copyright 2015 American Chemical Society.

mixed solution of pC7F15MAA and p(DDA/PtTPP) (95:5 w/w) in AK-225 and acetic acid (4:1 v/v) was cast on a glass substrate. The cast film was kept for at least one day at room temperature to remove solvents. The pC7F15MAA/p(DDA/PtTPP) film exhibited a water contact angle of 162°, indicating a superhydrophobic surface. More interestingly, the film showed great sensitivity to dissolved oxygen. (Figure 7) The luminescence intensity ratio at two oxygen concentrations reaches 126 ($=I_0/I_{40}$, corresponding to 0 and 40 mg L^{-1} dissolved oxygen concentrations). Conducting time-resolved luminescence lifetime measurements in water as well as in gas phase, we found that the equilibrium state is maintained between the entrapped gas and dissolved oxygen through the superhydrophobic surface of the pC7F15MAA/p(DDA/PtTPP) nanoparticle film. Table 2 summarizes the film-based sensor performance. For some of references, we refer to a recent review reported by Lo et al.[7] Our system has the highest oxygen sensitivity because of the porosity as well as superhydrophobic characteristics of amphiphilic fluorinated polymer nanoparticle films.

Table 2. Dissolved oxygen sensitivity.

Chromophore	Support matrix	K_{SV}	Sensitivity (I_0/I)	Ref.
$Ru(Ph_2phen)_3^{2+}$	TEOS/MTEOS	0.059 (not linear)	3.5 (0-40 ppm)	[8]
$Ru(dpp)_3^{2+}$	TMOS/DiMe-DMOS	0.14	16 (0-100 %)	[9]
PtOEP	Fluorinated co-polymer matrix	not linear	1.8 (0-40 mg L^{-1})	[10]
$Ru(Ph_2phen)_3Cl_2$	Fluropolymer matrix	0.22	6.5 (0-30 mg L^{-1})	[11]
$Ru(dpp)_3^{2+}$	TMOS/C8TMOS	0.11	10 (0-100 %)	[12]
$Ru(dpp)_3Cl_2$	TMOS/DiMe-DMOS	0.15	7.3 (0-40 mg L^{-1})	[13]
$Ru(dpp)_3Cl_2$	Porous plastic fiber	0.5523	8.7 (0-14 mg L^{-1})	[14]
$Ru(dpp)_3^{2+}$	TEOS/MTEOS	not linear	1.3 (0-45 mg L^{-1})	[15]
PtTFPP	TMOS/C8TMOS (silica nanoparticle)	4.18, 0.000005 (not linear)	117 (0-40 mg L^{-1})	[16]
p(DDA/PtTPP)	Amphiphilic fluorinated polymer	3.1	126 (0-40 mg L^{-1})	[4]

4. Conclusion

In conclusion, we described a unique bottom-up approach for creating amphiphilic fluorinated polymer nanoparticle films. Casting mixed solutions of amphiphilic fluorinated polymer, pC7F15MAA allows to prepare porous superhydrophobic surfaces. We addressed preparation conditions in terms of the mixing ratio of two solvents (acetic acid and AK-225) and solution concentration. We compared our sensing system for dissolved oxygen detection based on surface wettability and oxygen sensitive luminescence with other previous reports. Because of the good oxygen permeability as well as low surface tension of fluorinated side chains, the oxygen detection was successfully demonstrated with the excellent oxygen sensitivity in water (I_0/I_{40} = 126), which is the highest value reported so far. Other possibilities such as selective solvent extraction and wettability patterns, which are based on unique surface wettability of pC7F15MAA nanoparticle films will be clarified in the future.

Acknowledgements
This work was supported by a Grant-in-aid for Challenging Exploratory Research (15K13777). The work was also supported by the Nano-Macro Materials, Devices and System Research Alliance (MEXT). M. M. thanks the Asahi Glass Foundation for financial support.

References
[1] Caruso F, Caruso R A and Mohwald H 1998 *Science* **282** 1111-4
[2] Riess G 2003 *Prog. Polym. Sci.* **28** 1107-70
[3] Balazs A C, Emrick T and Russell T P 2006 *Science* **314** 1107-10
[4] Gao Y, Chen T, Yamamoto S, Miyashita T and Mitsuishi M 2015 *ACS Applied Materials & Interfaces* **7** 3468-72
[5] Mitsuishi M, Tanaka H, Obata M and Miyashita T 2010 *Langmuir* **26** 15117-20
[6] Obata M, Tanaka Y, Araki N, Hirohara S, Yano S, Mitsuo K, Asai K, Harada M, Kakuchi T and Ohtsuki C 2005 *J. Polym. Sci., Part A: Polym. Chem.* **43** 2997-3006
[7] Chu C-S, Lo Y-L and Sung T-W 2011 *Photonic Sens.* **1** 234-50
[8] McEvoy A K, McDonagh C M and MacCraith B D 1996 *Analyst* **121** 785-8
[9] Chen X, Zhong Z M, Li Z, Jiang Y Q, Wang X R and Wong K Y 2002 *Sens. Actuators B* **87** 233-8
[10] Gillanders R N, Tedford M C, Crilly P J and Bailey R T 2004 *Anal. Chim. Acta* **502** 1-6
[11] Gillanders R N, Tedford M C, Crilly P J and Bailey R T 2005 *Anal. Chim. Acta* **545** 189-94
[12] Tao Z Y, Tehan E C, Tang Y and Bright F V 2006 *Anal. Chem.* **78** 1939-45
[13] Pang H-L, Kwok N-Y, Chow L M-C, Yeung C-H, Wong K-Y, Chen X and Wang X 2007 *Sens. Actuators B* **123** 120-6
[14] Guo L, Ni Q, Li J, Zhang L, Lin X, Xie Z and Chen G 2008 *Talanta* **74** 1032-7
[15] Chu F, Yang J, Cai H, Qu R and Fang Z 2009 *Appl. Opt.* **48** 338-42
[16] Chu C S and Lo Y L 2010 *Sens. Actuators B* **151** 83-9

Conducting paper based sensor for cancer biomarker detection

Saurabh Kumar, Suveen Kumar, Chandra Mouli Pandey and Bansi D. Malhotra*

Nanobioelectronics Laboratory, Department of Biotechnology, Delhi Technological University, Shahbad Daulatpur, Delhi110042, India

***Correspondence:** bansi.malhotra@gmail.com

Abstract. We report results of studies relating to the fabrication of a paper based impedimetric biosensor using poly(3,4ethylenedioxythiophene):poly(4-styrene sulfonate) (PEDOT:PSS) modified Whatman paper. The layer- by- layer assembly of PEDOT:PSS deposited on the paper substrate results in the formation of a stable, conductive and homogenous film. The film has been characterized using Fourier transfer Infrared spectroscopy and scanning electron microscopy. This conducting paper platform has been used for the conjugation of anti-carcinoembryonic antigen (CEA) protein for quantitative estimation of CEA, a cancer biomarker that is frequently used in detection and monitoring of cancer. The results of the electrochemical impedimetric response studies indicate that the fabricated paper electrode can be used to estimate CEA in the range from 6-20 ngmL^{-1}, has sensitivity of 3.6 Ω mL ng^{-1} with a lower detection limit of 2.68 ng mL^{-1}.

1. Introduction

The availability of reliable diagnostic tools is an essential requirement in both the developing and developed world [1,2]. In this context there has been an emerging trend for evolution of new methods in nanotechnology, biotechnology and materials science for the development of point-of-care diagnostics (POC) devices. POC devices are analytical platforms that can be used to obtain rapid and accurate result at reasonable cost for biochemical analysis near the vicinity of patient [3,4].

Cancer is a serious concern and medical threat to the world because millions of people die due to uncontrolled cell growth every year. Carcinoembryonic antigen (CEA) is a glycoprotein involved in cell adhesion that is normally produced in gastrointestinal tissue and their level increases in serum of colorectal, lung and breast cancer. CEA determination is mostly used for diagnosis and to monitor patients who suffer from cancer which produce CEA [5-7]. Therefore detection of CEA is very important for diagnosis of cancer or more specifically monitoring of patient after and before getting chemotherapy, surgery, radiation or combination of all.

Paper is composed of cellulose fibre, which is a unit of the un-branched glucose. Due to flexibility, light weight, low cost, easily stored, transported, combustible, biocompatibility and biosafety management paper substrate can be advantageous for development of POC devices [8,9]. Conducting paper has been recently explored for developing innovative techniques for printed basic electronics components and inexpensive sensing kits [10,11]. *Kumar et. al.* fabricated conducting paper electrode by screen printing of silver paste over filter paper. These conducting paper electrodes were later used for electrochemical deposition of polyaniline to bind anti-sIL2Rα for detection of cancer [12]. *Ge et. al.* fabricated electrochemiluminescent immunodevice by chitosan grafted over screen printed conducting carbon paste on paper substrate and subsequent labelling of antibody with ruthenium complex for multiplexed detection of cancer biomarker [13]. These conducting polymers require a conducting substrate for electrochemical detection. Among the various polymer PEDOT: PSS has been found to have high ductility, thermal stability, easy handling of making film and ability to replace the traditionally used electrodes for flexible electronics [14,15].

In the present work a simple approach was used for the fabrication of conducting paper strips by progressively making a film of PEDOT:PSS (Poly(3,4ethylenedioxythiophene)/poly(4-styrene sulfonate)) directly over paper substrate. Electrochemical impedance spectroscopy (EIS) was used to study the interaction of CEA at the conducting paper electrode.

2. Experimental section

2.1.Material and Chemicals

Poly (3, 4-ethylenedioxythiophene)-poly(styrenesulfonate) 1.3 wt %, N-(3-dimethylaminopropyl)-N'-ethylcarbodiimide hydrochloride (EDC), N-hydroxysuccinimide (NHS), Carcinoembryonic Antibody Monoclonal (anti-CEA). Bovine serum albumin (BSA) and carcinoembryonic antigen (CEA) were purchased from Sigma Aldrich, India. Whatman Filter papers #1 was used as a substrate and all other chemicals were of analytical grade. The PEDOT:PSS solution was ultrasonically treated prior to being use. All other chemicals were of analytical grade and were used without further purification. Deionized water (Millipore) was used in all the buffer and solution preparations.

2.2.Fabrication of conducting and hydrophobic paper strips

Whatman filter paper #1 was cut into dimensions of 1cm × 3cm strips which were used as substrate. The paper strips treated with 2-3 drops of Tween 80 (detergent) in distilled water (DW) were washed with copious amount of water. The strips were further dipped into ethanol solution and sonicated for 15 min, followed by rinsing with DW and keep it in an oven at 60°C till dry. The paper strips were then dipped into PEDOT: PSS solutions for 10 min and then annealed at 110°C for 5 min. These consecutive cycles were repeated until desirable electrical parameter and integrity or mechanical strength of paper was obtained. The number of cycles corresponds to the number of PEDOT:PSS layer deposited on to the paper substrate. The color change from white to black clearly shows (Fig 1) the deposition of PEDOT:PSS and with increase in the deposition cycle (6 cycle) conducting paper strips was fabricated. These conducting paper strips were further treated with methanol at 100°C for 5 min to remove any artifacts from surface and to maintain the homogeneity on surface.

2.3.Sensor strips fabrication

1% (v/v) solution of (3-aminopropyl) triethoxysilane (APTES) solution (in toluene) was used to functionalize conducting paper strips. These modified conducting paper strips were immobilized with CEA (monoclonal) using EDC (0.2 M) as the coupling agent and NHS (0.05M) as activator (Scheme 1), for 2 h in dark. The modified paper strips was washed several time with phosphate buffer (PBS. 50 mM. pH 7.4, 0.9% NaCl) to remove any unbound antibodies. Bovine serum albumin (BSA) solution (2 mg/mL) was used to block non-specific active sites of the electrode. The BSA/anti-CEA/PEDOT:PSS paper electrodes were then washed with PBS and stored at 4°C, when not in use.

Journal of Physics: Conference Series **704** (2016) 012010 doi:10.1088/1742-6596/704/1/012010

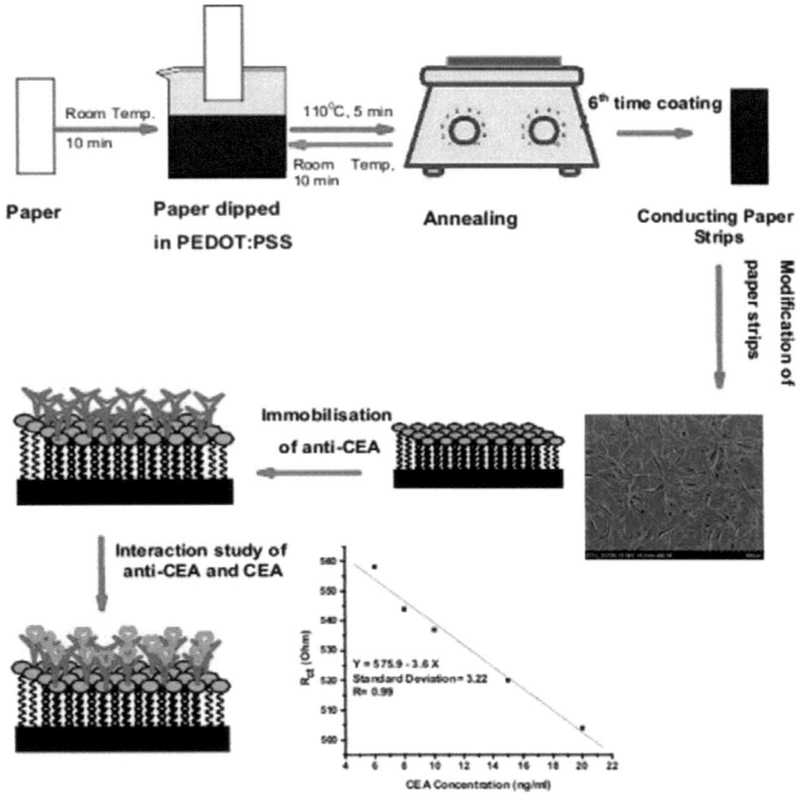

Scheme 1. Stratagem used for fabrication of Disposable sensor strips.

2.4.Measurement and characterizations

The electrochemical impedance studies (EIS) were performed using Autolab Potentiostat/Galvanostat (Metrohms, Netherlands) with frequency impedance analyzer. These experiments were carried out using a conventional three-electrode cell with disposable sensor strip as a working electrode, platinum as auxiliary electrode and Ag/AgCl as a reference electrode in phosphate buffer saline (PBS, 50 mM, pH 7.4) containing 5 mM $[Fe(CN)6]^{3-/4-}$at a 0.01V potential over the frequency range from 100KHz to 10Hz. Fourier transform infrared (FTIR) studies of conducting paper strips and APTES functionalized conducting paper strips were recorded by Thermo-Scientific Nicolet 380 spectrometer in reflectance mode in the region of 600-4000 cm^{-1}. The surface morphology of the conducting paper and APTES functionalized conducting paper was investigated by scanning electron microscope (Hitachi S-3700N). The contact angle measurements (DSA100, DSA/V 1.9, Kruss Germany) was done to study the surface properties of the modified electrode.

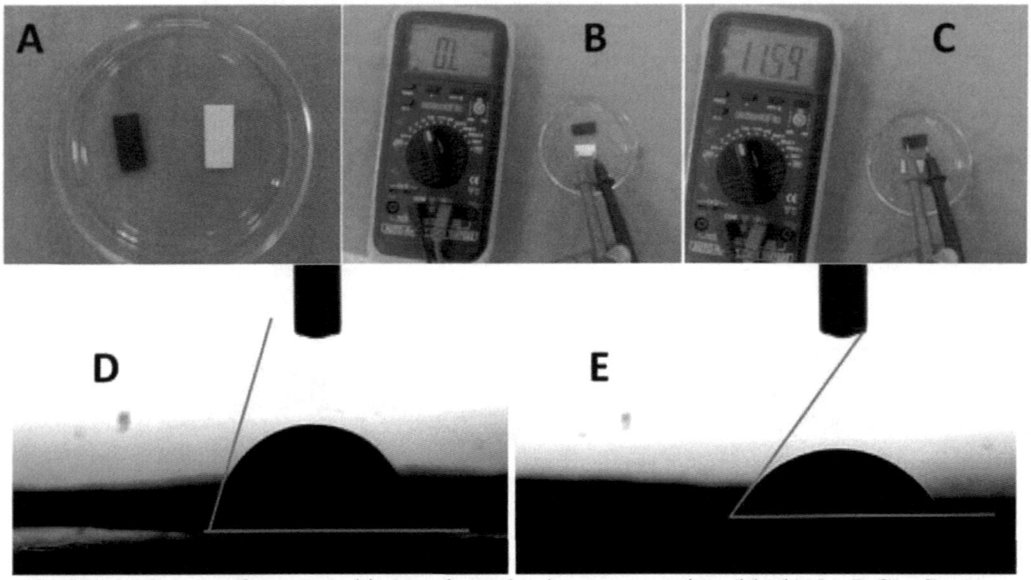

Figure 1. Optical image of paper (white) and conducting paper strips (black) [A,B,C]. Contact angle measurement of conducting paper strips [D] APTES functionalized Conducting paper Strips [E]

3. Results and discussion

The optimization of conducting paper strips was analyzed using EIS. Impedance measurements were done by applying sinusoidal AC voltage of different frequencies for interfacial studies of electrolyte and paper electrodes. EIS plot represented as a complex no. Z= Z' + jZ" where Z is the real part plotted as a X-axis and Z" is an imaginary part plotted as a Y-axis called Nyquist plot. In the Nyquist plot each point represents impedance at one frequency whereas a portion of semicircle observed at high modulation frequency describes the faradic electron transfer process between electrode and solution [16]. Data analysis after various stages of modification with best fitted equivalent electrical circuit model $R_s(R_pC_{dl})$ by fit and simulation software (NOVA 1.9), where Rs is solution phase resistance, R_{ct} is the charge transfer resistance and C_{dl} is double-layer capacitance that forms at the interface. The diameters of semicircle in Nyquist plot were utilized to observe the change in charge transfer resistant (R_{ct}) at electrode surface [17]. As a control experiment EIS studies for plane paper was also conducted. As the paper is dipped into electrolyte PBS (pH 7.4, containing 5 mM $[Fe(CN)6]^{3-/4-}$), an electrical contact is established due to hydrophilic nature of cellulose fibers. The Nyquist plot obtained for the paper shows a distorted semicircle indicating that paper behaves as a membrane and forms dielectric layer with leakage (data not shown) [16,18]. Controlling the noise by increasing the deposition cycle leads towards the hydrophobic nature of paper strips that restrict the wicking ability of paper and block the pores, which are most appropriate for measurement for the interfacial study between paper strip and solution. As the deposition cycle increases R_{ct} value decreases and heterogenous electron transfer rate constant increases (Table 1). Charge transfer resistance of conducting paper at each step controls the electron transfer kinetics of the redox probe $[Fe(CN)6]^{3-/4-}$. The heterogeneous electron transfer rate constant (K_{et}) between conducting paper electrode and the solution can be calculated using Eq(1) [19]:

$$\mathbf{K_{et} = RT/n^2F^2\ A\ R_{ct}\ [S] .... (1)}$$

where R is the gas constant, T is absolute temperature, F is Faraday constant, A is the electrode area (cm2), [S] is the concentration of redox probe (molcm3) and n is the number of transferred electrons per molecule of redox probe. We selected 6 times coated paper strips for further studies because there was no significance change in Rct value after further deposition (desirable electrical parameter obtained).

Table 1. Optimization of paper strips coated with PEDOT: PSS

Paper Strip Coating	R_{ct}	K_{ct}
4th times	92 K-Ohm	58 cm s^{-1}
5th times	24 K-Ohm	222 cm s^{-1}
6th times	15 K-Ohm	355 cm s^{-1}

3.1.Hydrophobic nature of paper

Paper is physically hydrophilic in nature because hydroxyl groups present in cellulose fiber have strong attraction with water and fluid flow through capillary action. Therefore the contact angle (CA) value of the paper approaches to zero. The wicking properties of paper are controlled by enclosing the channel of fibers by annealing with PEDOT: PSS. Conducting paper strip has a water contact angle of 77° confirming hydrophobic surface of conducting paper strips. The surface of conducting paper strips was functionalized by amine (-NH_2) group by treating it with 3-(aminopropyl)triethoxysilane and CA value decrease to 67° attributed to hydrophilic NH_2 group present in APTES (Fig 1).

3.2.Impedance study of conducting paper strips

In case of the conducting paper electrode, impedance at high frequencies under kinetic control and diffusion is negligible but at lower frequencies conducting paper electrode, distorted semicircle is observed because of the large pores and non-uniform layer (Fig 2A). Circuit was further improved by treatment of PEDOT: PSS coated paper with methanol. Treatment with methanol makes the film homogenous that reduces the surface- to- volume ratio of PEDOT: PSS coated paper .The charge transfer resistance increases up to 160KΩ (curve b) and distortion in the semicircle is reduced. Deposition of APTES at PEDOT:PSS coated paper strips was used for immobilization of biomolecule that acts as a molecular wire for faster transfer of charge. The decrease in the value of Rct after APTES deposition found as 75KΩ (curve c) is attributed to enhanced transfer of charge from solution to paper electrode due to polarization of terminal -NH_2 group of APTES, resulting in net positive charge on surface attracting the redox mediator $[Fe(CN)_6]^{-3/-4}$.

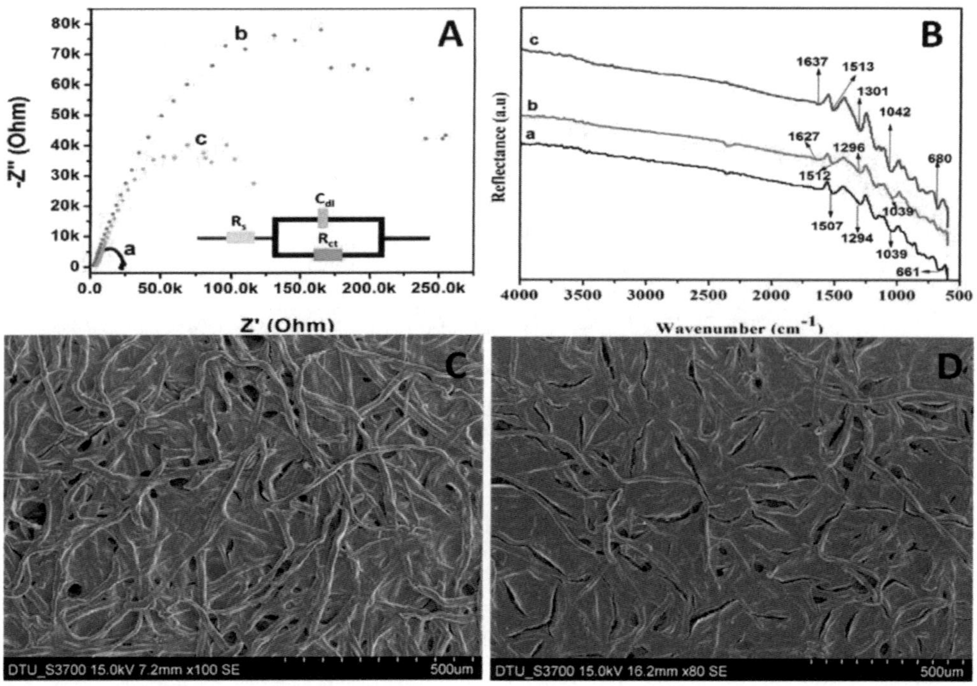

Figure 2. A) Electrochemical impedance spectra of a) Conducting paper strips b) Methanol treated conducting paper strips c) APTES functionalized surface, Inset showing equivalent circuit diagram. B) FTIR spectra of a) conducting paper strips b) Methanol treated conducting paper strips c) APTES functionalized surface. SEM image of C) conducting paper strips D) Methanol treated conducting strips.

3.3. FTIR studies

The FTIR spectra of fabricated paper based on PEDOT:PSS, methanol functionalized and APTMS modified PEDOT:PSS was carried out in the range 600-4000 cm^{-1} and is shown in Fig 2B. The characteristic FTIR band appears at 1507 cm^{-1} (C=C), 1294 cm^{-1} (C-C), 1039 cm^{-1} (S-O), and 661 cm^{-1} (C-S) corresponds to the thiophene ring in PEDOT:PSS [20]. The positions and relative bands of the transmittance peak of the PEDOT:PSS (curve a) and methanol functionalized PEDOT:PSS (curve b) are in agreement with literature. The IR spectra confirm that all regions have essentially the same chemical composition (curve a & b), where the most characteristic bands for PEDOT:PSS polymers can be clearly seen. The slight shifting of band at higher wavelength from 1507 cm^{-1} to 1512 cm^{-1} corresponds to C=C, which may perhaps be due to surface modification of PEDOT:PSS by methanol. The IR bands of PEDOT:PSS overlap with the IR band arises by APTMS modified PEDOT:PSS (curve c). The APTMS modified PEDOT:PSS shows the prominent peak at 680 and 833 cm^{-1} for Si-O symmetric stretching and bending modes, respectively. The band arises at 1042 cm^{-1} for (Si-O-Si) stretching and 1513 cm^{-1} belongs to NH_2 deformation of primary amine present in APTMS modified PEDOT:PSS paper [21].

3.4.Scanning Electron Microscopy measurement

Surface morphology of PEDOT:PSS and methanol treated PEDOT: PSS are further probed through SEM analysis. It is observed (Fig. 2C) that PEDOT:PSS conducting polymer is adsorbed at cellulose fibers wherein the deposition is clearly visible between cellulose fibers of paper. In contrast, surface morphology of methanol treated with PEDOT: PSS becomes more homogenous along with a few pores compared to the untreated PEDOT: PSS strips (Fig. 2D).

3.5.Application of disposable sensor strips

EIS is a sensitive technique that can be used to measure the interaction of biomolecule at the electrode surface. The apparent heterogeneous electron-transfer rate constants (K_{ct}) of disposable paper strips as found to be 1706 cm s^{-1} reveals improvement in electrochemical behavior owing to the immobilization of anti-CEA antibody. Several models of equivalent circuits were used to analyze the obtained plots, which are appropriate for a PEDOT: PSS film. After examining several possible equivalent circuit diagrams, the best fitting over the frequencies ranging from 100 KHz-10 Hz using an equivalent circuit diagram depicted in Fig. 2A (inset), was obtained.

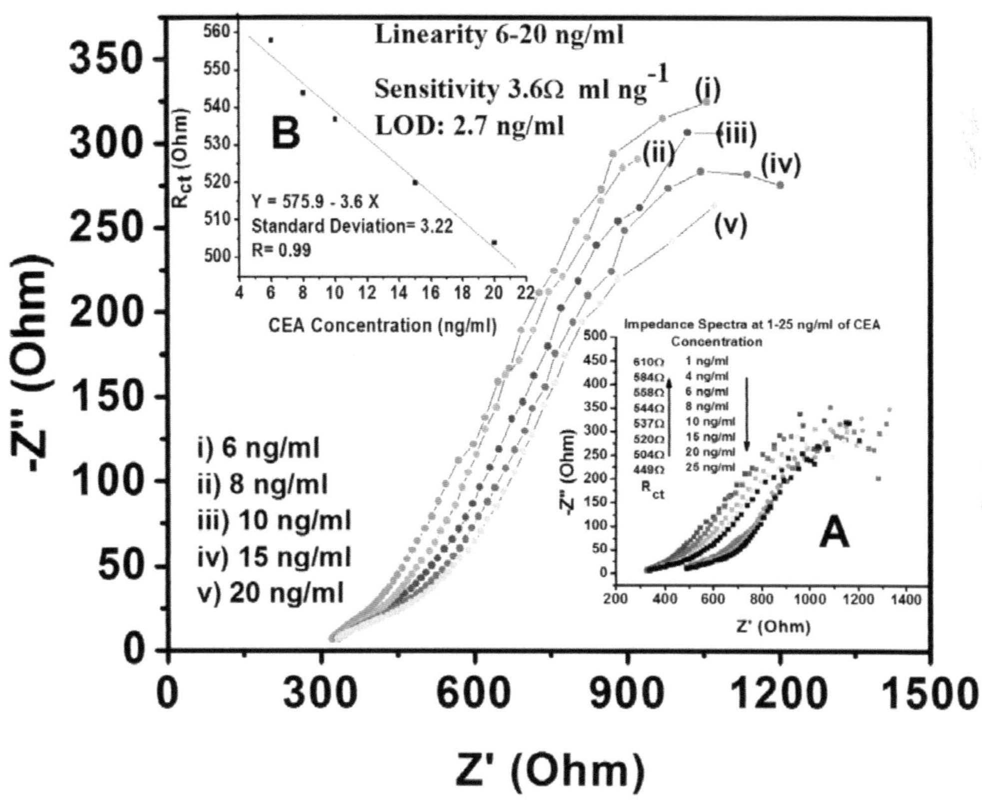

Figure 3. Impedance analysis of the disposable paper strips as a function of different concentration (1-25 ng/ml) of carcinoembryonic antigen (Inset A). Inset B showing a calibration plot between charge transfer resistance (Rct) and the CEA concentration (B).

The change in charge transfer resistance on addition of CEA from human fluids (human-CEA) was due to formation of antigen-antibody complex at the electrode surface. Immobilized antibodies at the electrode surface are worked with more CEA antigen as the concentration increases and correspondingly change in the parameter of impedance spectrum. It was observed that, there is a steady decrease in impedance as antigen concentration increases towards a concentration, 1-25 ng/ml (Fig 3). A linear relationship is found between R_{ct} and CEA concentration between 6-20 ng/ml with correlation coefficient of 0.9900, the relative change in Rct value observed at the different concentration of human-CEA and best fitted using equivalent circuit model. The linear regression equation is $\mathbf{R_{ct}(\Omega) = 576\ \Omega - 3.6\ \Omega\ ml\ ng^{-1} \times [CEA\ concentration\ (ng/ml)]}$ with a sensitivity of 3.6 Ω ml ng^{-1} and detection limit= 2.68 ng/ml calculated from the expression $3S_{y/x}$/*sensitivity*, where $S_{y/x}$ is standard deviation.

This immunosensor shows good sensor to sensor and batch to batch reproducibility and time taken for each measurement was ~5 min. It was observed that repeated runs on the same electrode gives reproducible results with RSD's less than 4% and the electrode-to-electrode reproducibility of responses have RSD less than 5% in a series of trials.

4.Conclusions

An electrochemical impedance based immunsensor has been fabricated using conducting poly (3,4ethylenedioxythiophene):poly(4-styrene sulfonate) modified paper. This paper based immunosensor has been used for selective detection of cancer biomarkers, where the inputs have been implied for the rapidity, sensitivity and stability of electrode. This simple approach does not require any highly sophisticated fabrication technique and conducting substrate such as gold, silver, ITO for impedance analysis. The results indicate promising application of this conducting paper in area of electronics, microfludics and energy storage device.

Acknoweledgements

We thank Vice Chancellor, Delhi Technological University (DTU) India for providing the facilities. Saurabh Kumar is thankful to DTU for the award of financial assistantship. C.M. Pandey acknowledges University Grant Commission, India for providing Dr. D. S. Kothari Postdoctoral fellowship (No.F.4-2/2006 (BSR)/CH/14-15/0164). We acknowledge Sandeep Mishra for SEM analysis.

References

[1] Yager P, Edwards T, Fu E, Helton K, Nelson K, Tam M R and Weigl B H 2006 *Nature* **442** 412-8
[2] Urdea M, Penny L A, Olmsted S S, Giovanni M Y, Kaspar P, Shepherd A, Wilson P, Dahl C A, Buchsbaum S and Moeller G 2006 *Nature* **444** 73-9
[3] Kumar S, Kumar S, Ali M, Anand P, Agrawal V V, John R, Maji S and Malhotra B D 2013 *Biotechnology journal* **8** 1267-79
[4] Kaushik A, Yndart A, Jayant R D, Sagar V, Atluri V, Bhansali S and Nair M 2015 *International journal of nanomedicine* **10** 677
[5] Benchimol S, Fuks A, Jothy S, Beauchemin N, Shirota K and Stanners C P 1989 *Cell* **57** 327-34
[6] Duffy M J 2001 *Clinical chemistry* **47** 624-30
[7] Aquino A, Formica V, Prete S P, Correale P P, Massara M C, Turriziani M, De Vecchis L and Bonmassar E 2004 *Pharmacological research* **49** 383-96
[8] Nery E W and Kubota L T 2013 *Analytical and bioanalytical chemistry* **405** 7573-95
[9] Kumar S, Kumar S, Srivastava S, Yadav B K, Lee S H, Sharma J G, Doval D C and Malhotra B D 2015 *Biosensors and Bioelectronics* **73** 114-122
[10] Kumar S, Willander M, Sharma J G and Malhotra B D 2015 *Journal of Materials Chemistry B* **3** 9305-14
[11] Jagadeesan K K, Kumar S and Sumana G 2012 *Electrochemistry Communications* **20** 71-4
[12] Kumar S, Jagadeesan K K, Joshi A G and Sumana G 2013 *RSC Advances* **3** 11846-53

[13] Ge L, Yan J, Song X, Yan M, Ge S and Yu J 2012 *Biomaterials* **33** 1024-31
[14] Lipomi D J, Lee J A, Vosgueritchian M, Tee B C-K, Bolander J A and Bao Z 2012 *Chemistry of Materials* **24** 373-82
[15] Groenendaal L, Jonas F, Freitag D, Pielartzik H and Reynolds J R 2000 *Advanced Materials* **12** 481-94
[16] Moisel M, de Mele M L and Müller W D 2008 *Advanced engineering materials* **10** B33-B46
[17] Daniels J S and Pourmand N 2007 *Electroanalysis* **19** 1239-57
[18] Atanasov V, Atanasova P P, Vockenroth I K, Knorr N and Köper I 2006 *Bioconjugate chemistry* **17** 631-7
[19] Bardea A, Katz E and Willner I 2000 *Electroanalysis* **12** 1097-106
[20] Ely F, Matsumoto A, Zoetebier B, Peressinotto V S, Hirata M K, de Sousa D A and Maciel R 2014 *Organic Electronics* **15** 1062-70
[21] Sharma A, Sumana G, Sapra S and Malhotra B D 2013 *Langmuir* **29** 8753-62

Nanomaterials based biosensors for cancer biomarker detection

Bansi D Malhotra*, Saurabh Kumar and Chandra Mouli Pandey

Nanobioelectronics Laboratory, Department of Biotechnology, Delhi Technological University, Delhi 110042, India

*Corresponding author: bansi.malhotra@gmail.com

Abstract. Biosensors have enormous potential to contribute to the evolution of new molecular diagnostic techniques for patients suffering with cancerous diseases. A major obstacle preventing faster development of biosensors pertains to the fact that cancer is a highly complex set of diseases. The oncologists currently rely on a few biomarkers and histological characterization of tumors. Some of the signatures include epigenetic and genetic markers, protein profiles, changes in gene expression, and post-translational modifications of proteins. These molecular signatures offer new opportunities for development of biosensors for cancer detection. In this context, conducting paper has recently been found to play an important role towards the fabrication of a biosensor for cancer biomarker detection. In this paper we will focus on results of some of the recent studies obtained in our laboratories relating to fabrication and application of nanomaterial modified paper based biosensors for cancer biomarker detection.

Keywords. Biosensor, Nanomaterials, Conducting paper, Cancer biomarker

1. Introduction

Cancer is currently the leading cause of morbidity and mortality worldwide and it is estimated that more than 11 million people are diagnosed with cancer every year and there will be 16 million new cases every year by 2020 [1, 2]. Cancer is defined as abnormal and uncontrolled cell growth due to an accumulation of specific genetic and epigenetic defects, which are either environmental or hereditary in origin [3]. The unregulated cell growth leads to the formation of a tumor mass that over time becomes independent of normal homeostatic checks and balances [4]. As the cancer progresses, the tumor begins to spread beyond the site of origin and metastasizes to other body organs and systems, making it incurable [3]. More than 200 distinct forms of cancer exist which include lung, prostate, breast, ovarian, hematologic, skin, and colon cancer, and leukemia etc [5]. Some cancers like stomach and cervical cancer are strongly associated with bacterial and viral infections ,respectively [6]. The conventional techniques for cancer detection such as centrifugation, chromatography, and fluorescence and magnetic-activated cell sorting rely on the expertise and subjective judgment of highly skilled personnel [7]. Diagnosing a cancer via cross sectional imaging (CT scan) and biopsy is an expensive and often uncomfortable approach for patients and yield substantial false-negative rates and a limited potential for early diagnosis of disease [8-10]. Therefore technologies to recognize and understand the signatures of normal cells and how these become cancerous, promise to provide important insights into the aetiology of cancer that can be useful for early detection, diagnosis, and treatment [11].

The detection of clinical biomarkers plays a crucial role in the early detection of a cancer, design of individual therapies, and to identify underlying processes involved in the disease [12]. Biomarkers are chemical substances related with the elevation of malignant tumors which are found in blood, urine, or body tissues [13]. They are normally produced directly by the embryonic tissue or tumor tissue [14]. Biomarkers indicate changes in the expression of a protein that is correlated to risk or progression of a disease or its response to treatment, and that can be measured in tissues or in the blood [14-15]. As a result, biomarkers can be specific cells, molecules or genes, gene products, enzymes or hormones [14]. An ideal cancer biomarker should have high clinical sensitivity and specificity, quick release in the blood enabling early diagnosis, capability to remain elevated for longer time in the blood, and ability to be assayed quantitatively [16-18]. More than 160 types of biomarkers are identified that are being used for the detection of cancer [19]. The expression of specific biomarkers and their accurate detection can be helpful in the diagnosis, staging and effective treatment of a cancer at an early stage. Various cancer markers are being widely used for diagnosing cancer, but carcinoembryonic antigen (CEA) is known to be a tumor marker associated with colon, lung, ovarian and breast cancer that are responsible for more than half of all cancer deaths each year [20]. CEA is a highly glycosylated cell surface glycoprotein, belonging to a group of substance known as the tumor-associated antigens. Hence, the monitoring of CEA level before and after cancer therapy facilitates early recognition of recurrences or detection of previously unremarked metastases. Determination of CEA in serum can be an interesting alternative for clinical diagnosis and monitoring of a cancer [21].

The quantification of biological or biochemical processes is of utmost importance for medical, biological and biotechnological applications [22]. However, converting the biological information to an easily processed electronic signal is challenging due to the complexity of connecting an electronic device directly to a biological environment. Electrochemical biosensors provide an attractive means to analyze the content of a biological sample due to direct conversion of a biological event to an electronic signal [22-23]. Over the past decades several sensing concepts and related devices have been developed. In this context, paper based electrochemical sensors are known to play an important role in on-going transition towards the POC diagnostic devices [24-25]. These papers based POC can be flat, folded, simple, low cost, flexible, and lightweight. Further, intelligent use of the polymers and nanomaterials (graphene, CNT) on paper may lead to enhanced performances with increased sensitivities and lowered detection limits of several orders of magnitudes. One advantage of utilizing these nanomaterials is the high specific surface thus already enabling the immobilization of an enhanced amount of bioreceptor units.

In this manuscript we discuss some of the research works reported on the development of nanomaterial modified paper substrate for cancer biomarker detection. Efforts have also been made to summarize some of the recent research work relating to CEA detection using CNT and graphene modified paper substrate.

2. Conventional method for CEA biomarker detection

Carcinoembryonic antigen (CEA) was described almost five decades ago by Gold and Freedman and was hypothesized to be an oncofetal antigen that is expressed in normal adult tissue as well [26]. Recently, CEA has been favored as a target antigen for primary and metastatic colorectal detection and other carcinomas of epithelial origin [27]. Owing to technological advancements various methods for CEA detection have been developed over the years ranging from immunohistopathology [28-29] to radioimmuno assays [30-31] (Enzyme linked immune sorbent assay-ELISA) [32-33] and reverse transcriptase polymerase chain reaction (RT-PCR) [34-35] to radioactive tracer fluoro-2-deoxy-D-glucose-Positron Emission Tomography (FDG-PET) [36-37] detection. Goldenberg et al reported a 3-step unlabeled antibody immunocytochemical staining method for CEA detection where horseradish peroxidase stain was used. The immunocytochemical reaction could be used to detect CEA concentrations above 0.7 μg/g in ethanol fixed and 3.0-5.0 μg/g in formalin fixed specimens. The CEA concentration values detected by this method are considerably high that indicates low sensitivity of the method [28]. Kleist et al used continuous cell line (HT29) from human carcinoma of the colon for synthesizing colon tumor antigen CEA which was detected by immunofluorescence technique. The weaker fluorescence of CEA indicated low sensitivity of the detection method [29]. Goldenberg et al reported a radioimmuno detection technique for CEA detection in which the prepared hyperimmune goat I_gG against CEA was radio labeled with ^{133}I (iodine radioactive isotope) [30]. The detection limit of this method was 2 cm and any tumor below 2 cm was undetectable which a major drawback of this method. RT-PCR based detection of CEA mRNA molecules in peripheral blood of 95 colorectal carcinoma (CRC) patients was carried out by Dong et al [34]. The sensitivity of the method achieved was 2 copies per tube that were high enough to determine very low levels of CEA in peripheral blood. Yonemura et al employed RT-PCR coupled with conventional cytological assay for determination of CEA mRNA in peritoneal washes of 230 patients of gastric cancer [35]. The sensitivity achieved by combining RT-PCR with cytological assay resulted in higher sensitivity over the conventional cytological assay alone.

Positron emission tomography (PET) can be employed to distinguish recurrent tumors from post therapeutic changes. PET is a functional imaging technique relying on physiological changes or metabolic functions for the detection of disease. Fluoro-2-deoxy-D-glucose (FDG) is a radioactive glucose analog used in medical imaging modality PET. The increased glucose metabolism of tumor cells is the base for using FDG-PET in oncology which makes it an important tool for quantitative analysis of malignant tumors. Flanagan et al used FDG-PET for investigating unexplained elevation of plasma CEA levels in 22 colorectal cancer patients post colorectal surgery. PET was more sensitive than conventional imaging methods with a positive predictive value of 89% (15 out of 17) in CRC patients [36]. ELISA has been considered as gold standard technique for detection of proteinaceous molecules and has been traditionally employed for tumor antigen, CEA detection [32]. Various modified ELISA techniques using gold nanoparticles have been used to further improve the sensitivity [33]. Inspite of the reliable detection, ELISA is a time consuming, complex and expensive detection method.

These conventional techniques hold advantage in providing low detection limits but are complex, time consuming and expensive. The requirement of trained personnel and expensive equipments as in case of FDG-PET is another major disadvantage of these techniques. They are not rapid and the diagnosis period ranges from a few hours to a few days making them laborious and time consuming. The success of these techniques has been marred by low sensitivity and specificity. Although a number of groups have modified the fundamental techniques for obtaining better results, like modified fluorescence RT-PCR, combining conventional cytology with RT-PCR, radioimmuno assays ,yet the achieved sensitivity and specificity of these techniques is low and calls for an urgent need for rapid, reliable, specific, and sensitive alternative technique for cancer detection.

3. Role of Biosensors in CEA detection

Early diagnosis of tumor biomarkers is critical for treatment of cancer. The urgent need for a rapid, reliable, specific and sensitive method for diagnosis of cancer markers at an early stage of disease has focused on point of care (POC) diagnostics. Biosensors have a number of potential advantages over other methods of cancer diagnosis, especially reduced assay time, portability, high sensitivity and selectivity, simplicity, miniaturized and flexibility. Biosensor-based diagnostics can assist cancer screening and improve the rates of earlier diagnosis and attendant improved forecast. This technology can be particularly useful for enhaned health care delivery in the public setting and to underserved diasporas. Biosensors have potential for multi-target analyses, automation, and cost effective testing [38].

Biosensor is a bioanalytical device, which converts a biological response into a measureable signal. Biosensor is used to estimate the concentration of desired substances and other parameters of biological interest without using the biological system directly. A biosensor consists of three major components such as a bio-recognition element for selective recognition of an analyte also known as bioreceptor, an immobilization matrix for the immobilization of a recognition biomolecule and a transducer for conversion of biochemical response into a measurable signal [39]. Bio-receptor and transducer are together referred to as a biosensing membrane. Based on the method of signal transduction biosensors may be classified into six basic groups, i.e. optical, mass, micromechanical, electrochemical, magnetic, and thermal sensors [40]. Among the various biosensors, electrochemical biosensors have received special attention as they allow high sensitivity, lower detection limits, automation, inexpensive testing, and development of disposable devices and methodologies capable of working with very small sample volumes [41].

3.1. Electrochemical Biosensor

An electrochemical biosensor is a small device that can be used for direct measurement of the analyte in the sample matrix. Ideally, such a device is capable of responding continuously and reversibly and does not perturb the sample. Electrochemical biosensors combine the analytical power of electrochemical techniques and the specificity of biological recognition processes. The most commonly used transducers in electrochemical biosensors are amperometric and potentiometric transducers. The analytical information in potentiometric devices is obtained by converting the biological response or biochemical reaction into a potential signal by the use of ionselective electrodes. Amperometric biosensors on the other hand monitor the current generated against applied constant potential by the reduction or oxidation of the electroactive species involved in the biorecognition process. Due to high sensitivity and wide linear range, ammperometric biosensors have attracted more attention. The ongoing research on new sensing concepts, combined with numerous technological advancements, has expanded horizons for extensive clinical applications of amperometric devices [42]. The enhanced sensitivity, specificity, simplicity, and inherent miniaturization of modern electrochemical bioassays allow them to compete with the most advanced optical protocols [41]. Feng et al reported an electrochemical biosensor for label free neoplastic cell detection using aptamer and functionalized graphene. The high binding affinity of the aptamer to the overexpressed nucleolin on the neoplastic cell surface enabled the electrochemical aptasensor to detect as low as thousand cells [43]. Chen et al developed a simple, label free electrochemical biosensor for oral cancer detection based on nuclease-assisted target recycling and DNAzyme for the detection of DNA species related to oral cancer in saliva [44].

3.2. Conventional electrodes used for cancer biomarker detection

For the fabrication of electrochemical biosensor for cancer biomarker detection various electrodes such as indium tin oxide, glassy carbon and gold electrodes have been used. Kumar et al used zirconia modified ITO electrode for oral cancer biomarker detection using cyclic voltammetry technique [45]. Zeng et al used nanostructured gold modified ITO electrode for CEA detection using electrochemical impedance spectroscopy [46]. Liu et al used gold electrode for fabrication of impedimetric biosensor for CEA detection [47]. In another study Huang et al used gold nanoparticle, carbon nanotube and chitosan modify glassy carbon electrode for CEA detection using differential pulse voltametery [48]. Lee et al reported a porous silicon antibody immunoassay platform for the detection of total human kallikrein 2, a protein involved the pathology of prostate cancer [49]. The major limitation of using the electrodes is they are costly, hard, brittle, non disposable and is difficult to functionalize. In addition to the aforementioned problems with the conventional electrodes, the requirement of very high temperatures for their processing and expertise makes them all the more undesirable. Therefore, in order to overcome these prevalent problems with the conventional electrodes, the attention has been drawn towards paper based electrode as a potential substrate.

Paper is becoming a suitable substrate for biosensing applications since it is simple, cost-effective, flexible, and are disposable. It is also beneficial in terms of cheap production, renewable raw materials and a porous structure useful for modification of paper substrate for sensing application [50]. Conducting paper (CP) has been considered as an efficient electrochemical platform for conduction of electronic signal generate during biosensing process [51]. For fabrication of CP both organic and inorganic materials can be used to make paper conducting. While inorganic materials often have a better electrical performance but the drawbacks are the high material cost, processing difficulties and cracks in film during bending/sintering. However, organic materials are preferred due to the mechanical flexibility, low cost, and simple processing. In this context, carbon nanomaterial, conducting polymer and their composite is promising candidate for fabricating a CP. The advantages of paper as a sensor substrate compared to the conventional electrodes are listed in Table 1.

Table 1. Comparison of the properties of the conventional electrodes with the paper electrode.

S.No	Properties	Conventional electrodes (Indium tin oxide and gold coated glass, Glassy carbon, Silicon)	Paper
1	Cost	Very high	Very low
2	Flexibility	No	Yes
3	Disposability	No	Yes
4	Biocompatibility	No	Yes
5	Modification/Functionalization	Difficult	Easy
6	Fluid flow	Forced	Capillary action
7	Surface by Volume ratio	Low	High
8	Fabrication and High throughput production	No	Yes

3.3. Graphene oxide integrated conducting paper for Cancer biomarker detection

Graphene is a flat monolayer of carbon atoms which are tightly packed into a two-dimensional (2D) honeycomb lattice. Graphene, has received much attention due to its unique physicochemical properties such as high surface area, high mechanical strength, and ease of functionalization, excellent conductivity and mass production [52]. Therefore, integration of graphene with conducting paper may perhaps lead to improved paper properties such as enhanced electrochemical activity, sensitivity and stability of paper sensor. Ruecha et al fabricated a paper based cholesterol biosensor with the help of electrospraying, wax printing and screen printing using graphene/polyvinylpyrrolidone/polyaniline nanocomposite for the detection of cholesterol [24]. The detection technique was amperometric and the fabricated electrode was found to be stable for about 2 weeks. Lu et al used graphene-gold nanoparticle for fabrication of paper based electrochemical DNA sensor by modifying paper substrate using Wax printing and screen printing [53]. In another work Xu et al reported an electrochemiluminescence sensor by modifying paper with poly(sodium 4-styrenesulfonate) functionalized graphene/nafion composite for the detection of DNA mismatches [54]. The electrode was fabricated with the help of photolithography and screen printing. Paper based chemiluminescence excited photoelectrochemical aptamer device for adenosine triphosphate measurement based on a gold coated paper electrode modified with a tin dioxide quantum dot/reduced graphene oxide nanocomposite has been fabricated by Wang et al [55]. Moreover, graphene oxide (GO) has been also incorporated in paper matrix for cancer biomarker detection. The dual-signal amplification technique was used by employing graphene oxide-chitosan/gold nanoparticles platform and [4,4′-(2,5-dimethoxy-1,4-phenylene)bis(ethyne-2,1-diyl) dibenzoic acid] (P-acid) functionalized nanoporous silver as signal amplification label [56]. The immunosensor was used to detect prostate specific antigen (PSA) and carcinoembryonic antigen (CEA) using electrochemical impedance spectroscopy. The linear detection range obtained for CEA was 0.001-10ng/mL with a LOD 0.8 pg/mL. The fabricated immunosensor was also found to be stable for about 4 weeks. Li et al detect CEA on paper platform by electrochemilumininescence based method. The materials used to fabricate the electrode were gold-chitosan hybrids and graphene quantum dots functionalized Au@Pt[57]. Further, Wu et al fabricated a paper based electrochemical immunodevice integrated with amplification-by-polymerization for the ultrasensitive multiplexed detection of cancer biomarkers (CEA, AFP, CA 125, CA 153). The material used in the electrode fabrication was GO and chitosan. The detection range of the fabricated sensor measured using differential pulse voltametery (DPV) was in the range, 0.01-100ng/mL [58].

The above discussed work require wax printing, screen printing and costly conducting ink (carbon paste, silver paste etc) and hold complicated design. To overcome this problem, we have recently fabricated a paper based label free electrochemical immunosensor based on PEDOT:PSS-RGO composite for the detection of CEA by chronoamperometric technique **(figure 1)** [59]. PEDOT:PSS-RGO modified conducting paper electrode are prepared by simple dip coating method, wherein the conductivity is further improved from $\sim 1.16\times10^{-4}$ Scm^{-1} up to $\sim 3.12\times10^{-2}$ Scm^{-1} on treatment with ethylene glycol. The observed significant increase in electrical conductivity is due to conformational rearrangement in the polymer and is due to strong non-covalent cooperative interaction between PEDOT and the cellulose molecules.

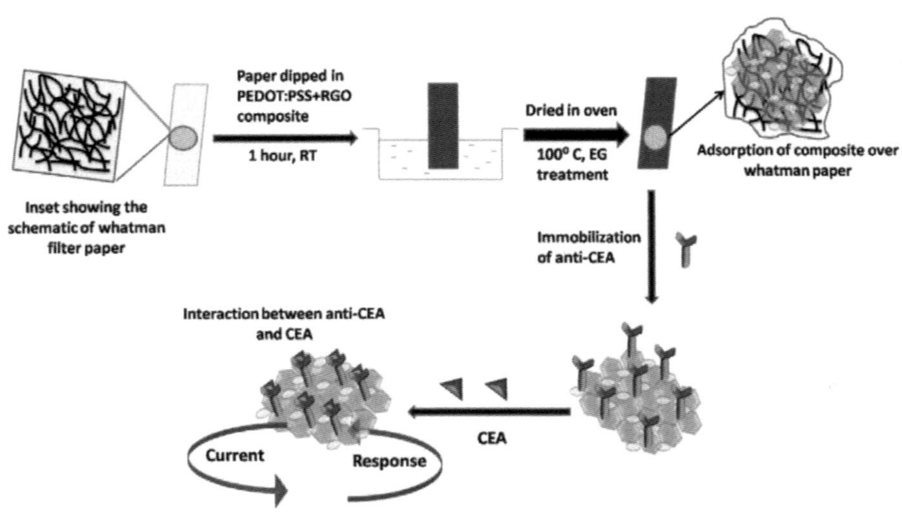

Figure 1. Schematic of PEDOT:PSS-RGO based paper biosensor for CEA detection[59]

Moreover the variation of relative conductivity of paper electrode with folding angle and folding cycle was studied **(figure 2)**. It is observed that 4% relative conductivity deviation during -180° to 180° folding angle and 20% conductivity deviation after 30 cycle of folding and unfolding (1 cycle = 360° rotation) was observed.

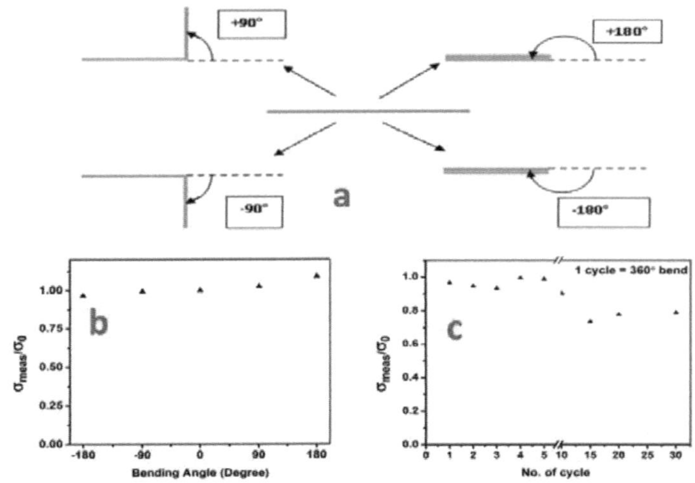

Figure 2. (a) Schematic diagram of PEDOT:PSS-RGO paper electrode folding at different deformation angle. (b) The ratio of measured conductivity (σ_{meas}) to initial conductivity (σ_o) of PEDOT:PSS-RGO based paper with respect to folding angle. (c) The ratio of measured conductivity (σ_{meas}) to initial conductivity (σ_o) of PEDOT:PSS-RGO based paper versus folding cycle (1 cycle= 360°)[59].

Integration of RGO in modified paper results in improved electrochemical activity and signal stability (Figure). Further this platform is used for the immobilization of monoclonal antibody against carcinoenmbryonic antigen (anti-CEA) (anti-CEA/RGO/EG@CP). The chronoamperometric studies were performed in PBS containing 5mM $[Fe(CN)_6]^{-3/-4}$ at 2 V. The sensing response of PEDOT:PSS-RGO composite based paper immunoelectrode as a function of CEA concentration **(figure 3)** revealed high sensitivity of 25.8 $mAng^{-1}mLcm^{-2}$ in the physiological range of 2–8 $ngmL^{-1}$ and good storage stability (21 days) for label free detection. The response of paper electrode is validated using CEA concentration of serum sample of cancer patient. This simple, low cost, flexible paper sensor can be decomposed by simple incineration and has immense potential as a smart medical diagnostic kit or point of care biosensor.

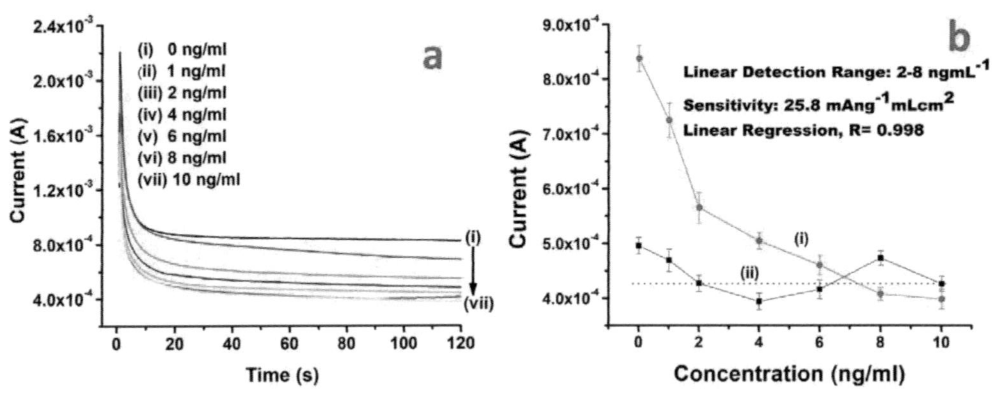

Figure 3. (a) Electrochemical response studies of anti-CEA immobilized PEDOT:PSS-RGO paper at different concentration of CEA. (b) Calibration plot between the magnitudes of current recorded and CEA concentration (curve i); control experiment in absence of antibody (curve ii)[59].

3.4. Carbon nanotube integrated conducting paper for cancer biomarker detection

Carbon nanotubes (CNTs) have been exploited for the development of electrochemical and biological sensors because of their excellent electrochemical properties, large surface area, ballistic electron transport and high mechanical strength. In addition to enhanced electrochemical reactivity, CNT-modified electrodes can be employed to immobilize biomolecules and to minimize surface fouling effects [60]. Moreover, CNT-based electrochemical transducers offer substantial improvements in the performance of amperometric enzyme electrodes, immunosensors and nucleic-acid sensing devices. Recent studies have demonstrated that CNT can be used to enhance the electrochemical reactivity of important biomolecules, and can be utilized to promote the electron-transfer reactions of proteins such as cytochrome c, ascorbic acid, xanthine oxidase, catalase, tryptophan and dopamine [61]. Further, CNT's exceptional properties such as small size, great strength, high electrical and thermal conductivity, and large specific surface area make them excellent amplification platforms to increase the number of signal-generating molecules [52].

Carbon nanotubes (CNTs) considered as a suitable candidate to be used in paper based sensing devices due to effective deposition characteristics using carbon nanotubes-based ink and excellent electronic transduction properties. For instance, Shim et al. took advantage of cotton yarns to build a SWCNT based chemiresistor for protein detection [62]. Wang et al modified filter paper with carbon nanotube and antibodies for amperometric detection of microcystin-LR toxin [63]. Similarly Pozuelo et al fabricated a conductive paper by simply painting paper filter with SWCNT ink. This platform was further utilized for human immunoglobulin G detection[64]. Zang et al reported electrochemical Immunoassay on 3D microfluidic paper-based device for detection of various cancer biomarker such as α-fetoprotein, carcinoma antigen 125, carcinoma antigen 199 and CEA [65]. For this purpose they modify paper surface with carbon paste, wax, carbon nanotubes, chitosan with the help of wax and screen printing technique. In another work, a battery-triggered microfluidic paper-based multiplex electrochemiluminescence immunodevice based on potential-resolution strategy was demonstrated for the diagnosis of four cancer markers (r-fetoprotein (AFP), CA 153, CA 199 and CEA [66]. For this purpose carbon nanotube, chitosan and glutardehyde modified paper zone were used for antibody immobilization which are connected with screen printed carbon working electrode and Ag/AgCl auxillary electrode through the paper channel. Further, Wang et al proposed a microfluidic paper-based analytical device (μPADs). A wax-patterned microfluidic paper-based three-dimensional electrochemical device (3D-μPED) was demonstrated based on the multi-walled carbon nanotubes (MWCNTs) modified μPAD. Using HRP-*O*-Phenylenediamine-H_2O_2 electrochemical system, a sandwich immunoassay on this 3D-μPED for sensitive diagnosis of two tumor markers simultaneously in real clinical serum samples was developed with a linear range of 0.001–75.0 U mL^{-1} for cancer antigen 125 and 0.05–50.0 ng mL^{-1} for CEA [67]. In this work different printing technique (screen printing, wax printing etc), costly conducting ink or sophisticated design were used for biomolecule detection.

We used composites of poly(3,4-ethylenedioxythiophene):poly(styrenesulfonate) (PEDOT:PSS) and carbon nanotube (CNT) for fabrication of the conducting paper by simple dip coating method[68]. Itwas found that the conductivity of formic acid treated paper increased by 2 orders of magnitude due to removal of non-conducting PSS molecule from conducting paper surface leading to conformational rearrangement. This fabricated paper is electrochemical active, flexible, efficient conductive and can be easily disposed off by simple incineration (**figure 4**). Further this platform was utilized for cancer biomarker (carcinoembryonic antigen, CEA) detection by immobilizing monoclonal antibody against CEA (anti-CEA). The PEDOT:PSS-CNT based electrochemical paper immunosensor showed sensitivity [7.8 $\mu A(ng/ml)^{-1}cm^2$] in a linear detection range of 2–15 $ngmL^{-1}$ (**figure 5**) and feasibility of paper electrode was also validated with CEA concentration in serum sample of cancer patient. It was observed that incorporation of carbon nanotubes improve heterogenous electron transfer rate constant and linear detection range of PEDOT:PSS-CNT based conducting paper compare to PEDOT:PSS-RGO based CP.

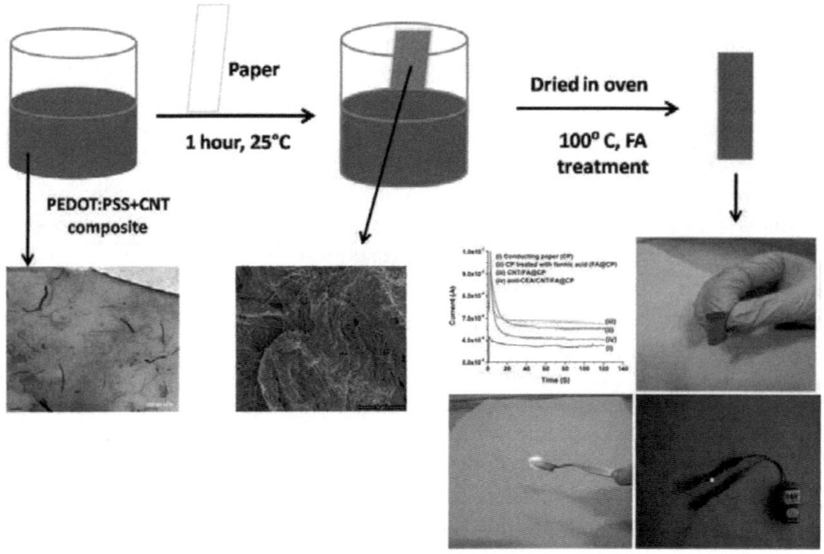

Figure 4. Fabrication process of PEDOT:PSS-CNT based paper and their properties (electrochemically active, flexible, easy to dispose off and optimum conductivity)[68].

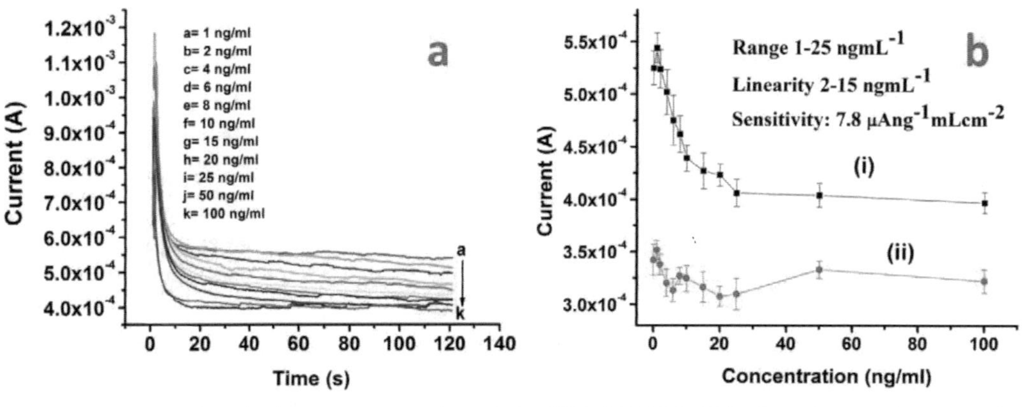

Figure 5. (a) Electrochemical response studies of the PEDOT:PSS-CNT based paper immunoelectrode obtained as a function of CEA concentration (1–100 ngml^{-1}) using chronoamperometry and (b) calibration plot between the magnitudes of current recorded and CEA concentration (curve i); control experiment in the absence of antibody (curve ii) [68].

4. Conclusions

We have discussed the prospects of CP based biosensors for cancer biomarker detection. Compared to the conventional electrode used for detection of CEA, CP based electrochemical biosensors are flexible, cost effective, light weight and can be easily disposed. The studies suggest that paper doped with MWCNT and graphene results in enhanced sensitivity, lower detection limit and wider linear detection range and have a great potential for early cancer detection. This nanomaterial modified paper based platform should be used in the development of medical diagnostics kits, flexible electronics and energy storages devices.

Acknowledgements
Saurabh Kumar is thankful to Delhi Technological University for the award of financial assistantship. C.M. Pandey acknowledges University Grant Commission, India for providing Dr. D. S. Kothari Postdoctoral fellowship (No.F.4-2/2006 (BSR)/CH/14-15/0164). The financial support received from Department of Science and Technology, India is acknowledged.

References

[1] http://www.who.int/mediacentre/factsheets/fs297/en/
[2] Thun M J, DeLancey J O, Center M M, Jemal A and Ward E M 2010 *Carcinogenesis* **31** 100-10
[3] Bohunicky B and Mousa S A 2011 *Science and Applications* **4** 1-10
[4] Hanahan D and Weinberg Robert A 2011 *Cell* **144** 646-74
[5] Rasooly A and Jacobson J 2006 *Biosens. Bioelectron.* **21** 1851-8
[6] Selgrad M, Malfertheiner P, Fini L, Goel A, Boland C R and Ricciardiello L 2008 *J. Cell. Physiol.* **216** 378-88
[7] Chen J, Li J and Sun Y 2012 *Lab on a Chip* **12** 1753-67
[8] Krebs M G, Hou J-M, Ward T H, Blackhall F H and Dive C 2010 *Ther. Adv. Med. Oncol.* **2** 351-65
[9] Hurt E and Farrar W 2008 *Mol Interv* **8** 140 - 2
[10] http://www.who.int/mediacentre/factsheets/fs297/en/
[11] Mehta S, Shelling A, Muthukaruppan A, Lasham A, Blenkiron C, Laking G and Print C 2010 *Ther. Adv. Med. Oncol.* **2** 125-48
[12] Madu C O and Lu Y 2010 *J. Cancer* **1** 150-77
[13] De Palma M and Hanahan D 2012 *Mol. Oncol.* **6** 111-27
[14] Henry N L and Hayes D F 2012 *Mol. Oncol.* **6** 140-6
[15] Preedy V R and Patel V B 2015 *Biomarkers in Cancer*: Springer
[16] Hori S S and Gambhir S S 2011 *Sci. Transl. Med.* **3** 109ra16-ra16
[17] Wu L and Qu X 2015 *Chem. Soc. Rev.* **44** 2963-97
[18] Gillette M A and Carr S A 2013 *Nat. Methods* **10** 28-34
[19] Kumar S, Kumar S, Ali M, Anand P, Agrawal V V, John R, Maji S and Malhotra B D 2013 *Biotechnology journal* **8** 1267-79
[20] Banin Hirata B K, Oda J M M, Losi Guembarovski R, Ariza C B, Oliveira C E C d and Watanabe M A E 2014 *Dis. Markers* **2014** 12
[21] Kobayashi E, Ueda Y, Matsuzaki S, Yokoyama T, Kimura T, Yoshino K, Fujita M, Kimura T and Enomoto T 2012 *Cancer Epidemiology Biomarkers & Prevention* **21** 1902-12
[22] Grieshaber D, MacKenzie R, Vörös J and Reimhult E 2008 *Sensors (Basel, Switzerland)* **8** 1400-58
[23] Frew J E and Hill H A O 1987 *Anal. Chem.* **59** 933A-44A
[24] Ruecha N, Rangkupan R, Rodthongkum N and Chailapakul O 2014 *Biosensors and Bioelectronics* **52** 13-9
[25] Liu B, Du D, Hua X, Yu X-Y and Lin Y 2014 *Electroanalysis* **26** 1214-23
[26] Gold P and Freedman S O 1965 *The Journal of Experimental Medicine* **122** 467-81
[27] Hammarström S 1999 *Semin. Cancer Biol.* **9** 67-81
[28] Goldenberg D M, Sharkey R M and Primus F J 1978 *Cancer* **42** 1546-53
[29] Von Kleist S, Chany E, Burtin P, King M and Fogh J 1975 *J. Natl. Cancer Inst.* **55** 555-60
[30] Goldenberg D M, Kim E E, DeLand F H, Bennett S and Primus F J 1980 *Cancer Res.* **40** 2984-92
[31] Thomson D, Krupey J, Freedman S and Gold P 1969 *Proceedings of the National Academy of Sciences* **64** 161-7
[32] Lai R-S, Hsu H-K, Lu J-Y, Ger L-P and Lai N-S 1996 *CHEST Journal* **109** 995-1000
[33] Zhou F, Wang M, Yuan L, Cheng Z, Wu Z and Chen H 2012 *Analyst* **137** 1779-84

[34] Xu D, Li X-f, Zheng S and Jiang W-z 2006 *Journal of Zhejiang University Science B* **7** 445-51
[35] Yonemura Y, Endou Y, Fujimura T, Fushida S, Bandou E, Kinoshita K, Sugiyama K, Sawa T, Kim B S and Sasaki T 2001 *ANZ J. Surg.* **71** 521-8
[36] Flanagan F L, Dehdashti F, Ogunbiyi O A, Kodner I J and Siegel B A 1998 *Ann. Surg.* **227** 319
[37] Liu C-S, Shen Y-Y, Lin C-C, Yen R-F and Kao C-H 2002 *Jap. J. Clin. Oncol.* **32** 244-7
[38] Tothill I E 2009 *Semin. Cell Dev. Biol.*: Elsevier pp 55-62
[39] Gerard M, Chaubey A and Malhotra B 2002 *Biosens. Bioelectron.* **17** 345-59
[40] Ivnitski D, Abdel-Hamid I, Atanasov P and Wilkins E 1999 *Biosens. Bioelectron.* **14** 599-624
[41] Wang J 2006 *Biosens. Bioelectron.* **21** 1887-92
[42] Wang J 1999 *J. Pharm. Biomed. Anal.* **19** 47-53
[43] Feng L, Chen Y, Ren J and Qu X 2011 *Biomaterials* **32** 2930-7
[44] Chen J, Zhang J, Guo Y, Li J, Fu F, Yang H-H and Chen G 2011 *Chem. Commun.* **47** 8004-6
[45] Kumar S, Kumar S, Tiwari S, Srivastava S, Srivastava M, Yadav B K, Kumar S, Tran T T, Dewan A K and Mulchandani A 2015 *Advanced Science* **2**
[46] Zeng H, Agyapong D A Y, Li C, Zhao R, Yang H, Wu C, Jiang Y and Liu Y 2015 *Sensors and Actuators B: Chemical* **221** 22-7
[47] Liu Z and Ma Z 2013 *Biosens. Bioelectron.* **46** 1-7
[48] Huang K-J, Niu D-J, Xie W-Z and Wang W 2010 *Anal. Chim. Acta* **659** 102-8
[49] Lee S W, Hosokawa K, Kim S, Jeong O C, Lilja H, Laurell T and Maeda M 2015 *Sensors* **15** 11972-87
[50] Nery E W and Kubota L T 2013 *Anal. Bioanal. Chem.* **405** 7573-95
[51] Desmet C, Marquette C A, Blum L J and Doumèche B 2016 *Biosensors and Bioelectronics* **76** 145-63
[52] Malhotra B D, Srivastava S and Augustine S 2015 *MRS Proceedings*: Cambridge Univ Press pp mrsf14-1725-i05-02
[53] Lu J, Ge S, Ge L, Yan M and Yu J 2012 *Electrochimica Acta* **80** 334-41
[54] Xu Y, Lou B, Lv Z, Zhou Z, Zhang L and Wang E 2013 *Analytica chimica acta* **763** 20-7
[55] Wang Y, Xu J, Ma C, Li S, Yu J, Ge S and Yan M 2014 *Journal of Materials Chemistry B* **2** 3462-8
[56] Li W, Li M, Ge S, Yan M, Huang J and Yu J 2013 *Analytica chimica acta* **767** 66-74
[57] Li L, Li W, Ma C, Yang H, Ge S and Yu J 2014 *Sensors and Actuators B: Chemical* **202** 314-22
[58] Wu Y, Xue P, Hui K M and Kang Y 2014 *Biosensors and Bioelectronics* **52** 180-7
[59] Kumar S, Kumar S, Srivastava S, Yadav B K, Lee S H, Sharma J G, Doval D C and Malhotra B D 2015 *Biosensors and Bioelectronics* **73** 114-122
[60] Singh C, Srivastava S, Ali M A, Gupta T K, Sumana G, Srivastava A, Mathur R and Malhotra B D 2013 *Sensors Actuators B: Chem.* **185** 258-64
[61] Lin Y, Yantasee W and Wang J 2005 *Front. Biosci* **10** 582
[62] Shim B S, Chen W, Doty C, Xu C and Kotov N A 2008 *Nano letters* **8** 4151-7
[63] Wang L, Chen W, Xu D, Shim B S, Zhu Y, Sun F, Liu L, Peng C, Jin Z and Xu C 2009 *Nano Lett.* **9** 4147-52
[64] Pozuelo M, Blondeau P, Novell M, Andrade F J, Rius F X and Riu J 2013 *Biosensors and Bioelectronics* **49** 462-5
[65] Zang D, Ge L, Yan M, Song X and Yu J 2012 *Chemical Communications* **48** 4683-5
[66] Wang S, Ge L, Zhang Y, Song X, Li N, Ge S and Yu J 2012 *Lab on a Chip* **12** 4489-98
[67] Wang P, Ge L, Yan M, Song X, Ge S and Yu J 2012 *Biosensors and Bioelectronics* **32** 238-43
[68] Kumar S, Willander M, Sharma J G and Malhotra B D 2015 *Journal of Materials Chemistry B* **3** 9305-14

IJEGMBE 2015 IOP Publishing
Journal of Physics: Conference Series **704** (2016) 012012 doi:10.1088/1742-6596/704/1/012012

Photophysical investigations of squaraine and cyanine dyes and their interaction with bovine serum albumin

M Saikiran[1], D Sato[1], S S Pandey[1] and T Kato[1]

[1] Graduate School of Life Science and Systems Engineering, Kyushu Institute of Technology, 2-4 Hibikino, Wakamatsu, Kitakyushu 808-0196, Japan

e-mail:msaikiran87@gmail.com / shyam@life.kyutech.ac.jp

Abstract: A model far-red sensitive symmetrical squaraine dye (SQ-3) and unsymmetrical near infra-red sensitive cyanine dye (UCD-1) bearing direct –COOH functionalized indole ring were synthesized, characterized and subjected to photophysical investigations including their interaction with bovine serum albumin (BSA) as a model protein in phosphate buffer solution (PBS). Both of the dyes exhibit strong interaction with BSA in phosphate buffer with high apparent binding constant. A judicious tuning of hydrophobic main backbone with reactive functionality for associative interaction with active site of BSA has been found to be necessary for BSA detection in PBS.

1. Introduction

Biomedical diagnosis is rapidly growing area in order to provide high quality of life by early disease diagnosis. In this context, optical imaging has emerged as one of the potential candidates due to its low cost, low-energy radiation, high sensitivity and non-invasive nature [1]. In line with fast growing imaging technology, near infrared (NIR) fluorescence probes have shown their potential towards non-invasive *in vivo* imaging owing to their low tissue auto-fluorescence and deep tissue penetration [2]. They have generated considerable research interest in several areas of the biomedical field, including vascular mapping, tissue perfusion, inflammation monitoring and tumor diagnosis. Squaraine dyes belong to interesting organic chromophores consisting of a central four-membered ring-based core and a resonance stabilized zwitterionic structure. The oxocyclobutenolate core is linked by aromatic or heterocyclic components at both ends generating donor–acceptor–donor motif. Squaraine dyes were studied extensively in numerous areas, ranging from synthesis and mechanisms to physical and photophysical properties [3-4]. The absorption and emission behaviour of these dyes can be tuned in the visible and IR wavelength region by judicial molecular design making them suitable for fluorescence imaging since they are outside the self-absorption and auto-fluorescence regions of biological matrices [5]. These unique chemical and physical properties of squaraine molecules have led to studies of their application as near-IR fluorescent chromophores and environmental sensors [6], for bio imaging and biochemical labelling [7-8]. Cyanine dyes are another most extensively investigated organic dyes which are composed of two nitrogen-containing heterocycles as charged chromophores, each conjugated to the end of a polymethine that comprises an odd number of carbons [9]. Cyanine dyes find their versatile applications in the area of fluorescent probes for bio-labelling [10], analyte responsive fluorescent probes [11] and in optoelectronics [12]. Wide range fluorescence spectral tunability and good fluorescent quantum yield enable the cyanine dyes to detect very low concentrations of analytes [13]. Recent study on unsymmetrical cyanine dyes have gained good attention owing to their excellence nucleic acid staining properties [14].

Photophysical properties such as tuneable absorption and emission in the far-red to NIR region, high molar extinction coefficient, considerable quantum yields and photo-stabilities. Squaraine and cyanine dyes are of utmost interest as fluorescent labels/probes [15-16], biomedical assays [17] and imaging applications [18]. Due to the presence of negatively charged oxygen, zwitterionic squaraine fluorophores are considered more polar molecules as compared to open chain cationic cyanine dyes. Therefore, the quantum yield and fluorescence lifetime of squaraine dyes are considerably decreased, but the increase in quantum yield and lifetime upon covalent or non-covalent binding to most widely investigated model protein bovine serum albumin (BSA) has been reported earlier [19]. This paper describes the detailed synthesis and photophysical characterization of representative NIR cyanine (UCD-1) and far-red sensitive squaraine (SQ-3) dyes. In order to explore their potential application as fluorescent probes to sense protein in phosphate buffer solution (PBS), these dyes were subjected to study their interaction using BSA as a model protein. In order to enable covalent coupling with biomolecules such as peptides, oligonucleotides and proteins, the fluorescent probes should contain reactive functional groups. Therefore, we have chosen these type of molecules where direct ring functionalized carboxy groups which enables covalent attachment either to amino or thiol functionalities of biomolecules.

2. Materials and Methods

All the chemicals for synthesis and solvents are of analytical or spectroscopic grade and used as received without further purification. Unsymmetrical cyanine dye (UCD-1), symmetrical squaraine dye (SQ-3) and their intermediates under investigation (scheme 1) were analysed by MALDI-TOF-mass / FAB-mass spectrometry in positive ion monitoring mode for structural elucidation. Electronic absorption and fluorescence emission spectroscopic investigation in solution phase were executed using UV-visible and fluorescence spectrophotometers. In order to investigate, protein dye interactions, the PBS/Cyanine and PBS/squaraine dye solutions (2μM) were prepared by addition of 100 μl of 0.1 M dye solution in DMF to the different concentrations of PBS/BSA solutions (0-10 μM). The final solutions were gently stirred for 1 hour at room temperature. In order to compare the dye-protein interactions quantitatively, apparent binding constant (K_a) was also calculated from fluorescence titration of dyes at 25°C and at a constant dye concentration by dilution of the initial BSA concentration. Considering BSA/dye association in 1:1 ratio, K_a was calculated using the equation 1 [2, 20].

$$\frac{1}{(F_x - F_0)} = \frac{1}{(F_\infty - F_0)} + \frac{1}{K_a[BSA]}\frac{1}{(F_\infty | F_0)} \quad (1)$$

Where, F_0, F_X, and F_∞ are the fluorescence intensities of dyes in the absence of BSA, in the presence of a certain amount of BSA and at a concentration of complete interaction, respectively, while [BSA] is the protein concentration. Equation (1) can be rewritten as

$$\frac{(F_\infty - F_0)}{(F_x - F_0)} = 1 + \frac{1}{K_a[BSA]} \quad (2)$$

The binding constant values (K_a) between the BSA and the dyes thus be easily calculated from the slopes of the corresponding plots $(F_\infty - F_0)/(F_X - F_0)$ as a function of $[BSA]^{-1}$ as per the equation (2).

2.1. Synthesis of Squaraine and cyanine dyes and dye intermediates

Model far-red sensitive symmetrical squaraine dye (SQ-3) and unsymmetrical cyanine dye (UCD-1) has been synthesized as per the scheme-1 and reported literature procedures [21-26].

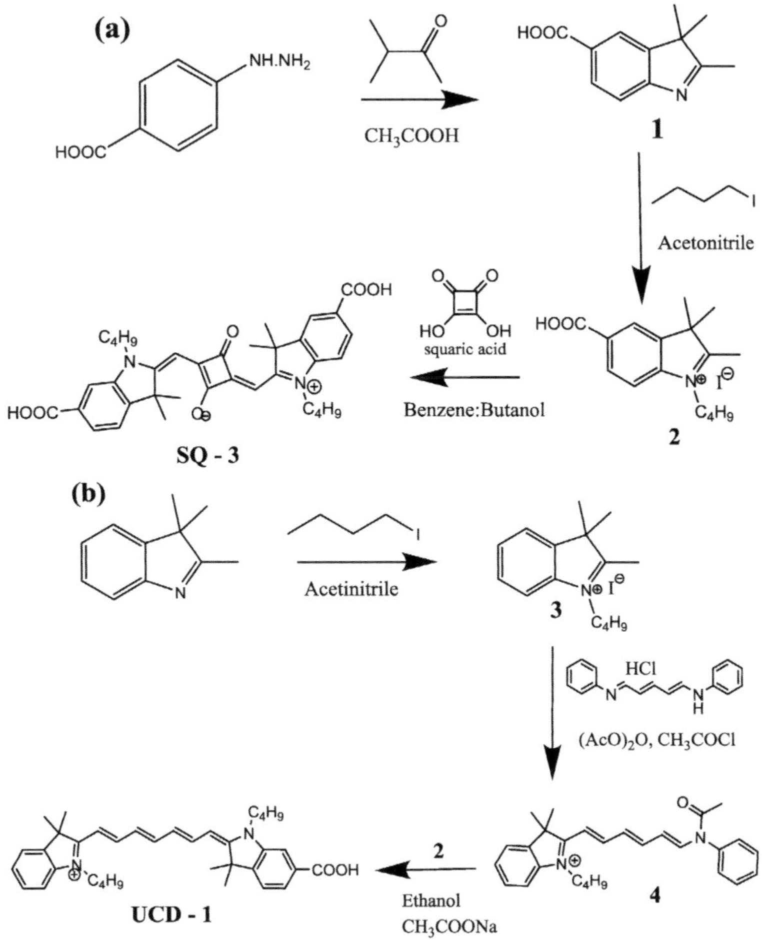

Scheme 1: Synthetic scheme for symmetrical squaraine (SQ-3) dye (a) and unsymmetrical cyanine (UCD -1) dye (b) used for present investigation.

2.1.1. Synthesis of 2,3,3-trimethyl-3H-indole-5-carboxylic acid (1): In a round bottom flask fitted with condenser and N_2 purging, 4-hydrazinobenzoic acid (5.0 g; 32.85 mmol), glacial acetic acid (100 ml), sodium acetate (5.5 g; 67 mmol) and 3-methyl-2-butanone (4.45 g; 51.5 mmol) were added. Reaction mixture was refluxed at 120 °C for 8 h leading to brown suspension. Upon the completion of reaction as monitored by TLC, acetic acid was evaporated followed by addition of 9:1 water methanol mixture on ice-bath leading to precipitation. Residue was filtered and dried giving 3.7 g of titled compound as of white powder in 56% yield. HPLC analysis of product suggests that compound was 100% pure. MALDI-TOF – mass (measured 205.2 $[M+2]^+$; calculated 203.24).

2.1.2. Synthesis of 1-butyl-5-carboxy-2,3,3-trimethyl-3H-indol-1-ium (2): In a round bottom flask one equivalent of (**1**) and 1-Iodobutane (3 eq.) were dissolved in dehydrated acetonitrile and the reaction was refluxed for overnight to give the corresponding alkyl-3H-indolium iodide. After completion of the reaction as monitored by TLC, solvent was evaporated and the crude product was washed with ample diethyl ether giving the titled compounds in 64% yield having 98% purity as conformed by HPLC. MALDI-TOF – mass (measured 261.13 $[M+H]^+$; calculated 260.36).

2.1.3. Synthesis of 1-butyl-2,3,3-trimethyl-3H-indol-1-ium iodide (3): In a round bottom flask, one equivalent of commercially available 2,3,3-trimethyl indole and 1-Iodobutane (3 eq.) were dissolved in dehydrated acetonitrile and the reaction was refluxed for overnight to give the corresponding alkyl-3H-indolium iodide. After completion of the reaction as monitored by TLC, solvent was evaporated and the crude product was washed with ample diethyl ether giving brown solid in 95% yield having 98% purity as conformed by HPLC. MALDI TOF – mass (measured 217 $[M+H]^+$; calculated 216).

2.1.4. Synthesis of Hemicyanine (4): In a round bottom flask one equivalent of alkyl-3H-indolium iodide (**3**) and glutaconaldehyde dianil monohydrochloride, along with catalytic amount of acetyl chloride were dissolved in acetic anhydride. The reaction mixture was refluxed at 140^0C for 2 hours. After the completion of reaction as monitored by TLC, the reaction mixture was poured on to crushed ice to precipitate the desired compound as black solid. This was filtered, dried and purified by silica gel column chromatography (Ethyl acetate: Hexane = 1:1). The pure compound after column chromatography was obtained as green solid in 60% yield. MALDI-TOF – mass (measured 414.33 $[M+H]^+$; calculated 413.58).

2.1.5. Synthesis of symmetrical Squaraine dye (SQ-3): The dye (**SQ-3**) was synthesized using two equivalents of carboxy functionalized trimethyl indolium iodide salt (**2**) and squaric acid (1 equiv.) in 1-butanol: toluene mixture (1:1, v/v). Reaction mixture was refluxed for overnight using Dean–Stark trap. After completion of reaction, reaction mixture was cooled, solvent was evaporated and product was purified by silica gel column chromatography using chloroform: methanol as eluting solvent. The physical and spectroscopic data of symmetrical SQ-3 dye are as follows; Yield 64% and HPLC purity 98%. MALDI-TOF-mass (calculated 596.29 and observed 597.25 $[M+H]^+$) which confirms the successful synthesis of dye.

2.1.6. Synthesis of Unsymmetrical Cyanine dye (UCD-1): In a round bottom flask one equivalent of hemi cyanine (**4**) was dissolved in 1-butanol: toluene (1:1, v/v). To the above solution one equivalent of Compound **2** and 2 equivalents of sodium acetate were added. The reaction mixture was refluxed at 95^0C for 2 hours. Upon the completion of reaction as monitored by TLC, the solvent was evaporated under reduced pressure and was purified by column chromatography on silica gel to afford titled compound as blue green solid in 43% yield having 95% purity as conformed by HPLC. HRMS (calculated 537.3476 and observed 537.3453 $[M]^+$).

3. Results and Discussion

3.1. *Photophysical Characterization*

Far-red to NIR sensitive dyes SQ-3 and UCD-1 after their successful synthesis and purification were subjected to spectroscopic investigations pertaining to the electronic absorption and fluorescence emission spectroscopy. Figure 1 depicts the electronic absorption and fluorescence emission spectra of both of the dyes UCD-1 and SQ-3 in dimethylformamide (DMF) solution. Characteristic optical parameters for these dyes deduced from Fig. 1 have also been summarized in the table 1. It can be seen from Fig. 1 that both of the dyes SQ-3 and UCD-1 exhibit very sharp and intense light absorption in far-red to NIR wavelength region. Absorption maxima (λ_{max}) of cyanine dye UCD-1 is located at 764 nm with molar extinction coefficient value (ε) (1×10^5 dm^3 M^{-1} cm^{-1}) which is associated with the π-π^* electronic transitions. In the case of squaraine dye SQ -3 ($\varepsilon = 2 \times 10^5$ dm^3 M^{-1} cm^{-1}) observed λ_{max}

is at 650 nm which relatively blue shifted. This is attributed to the reduced extent of p-conjugation in SQ-3 as compared to that of UCD-1. The fluorescent emission spectra for each of the dyes were measured using λ_{max} as excitation wavelength. The emission peaks for UCD-1 and SQ-3 has been observed at 788 nm and 662 nm, respectively with a Stoke shift of 24 nm and 12 nm. This lower Stoke shift for SQ-3 in DMF solution depicts the enhanced rigidity of squaraine dye as compared to that of cyanine dye.

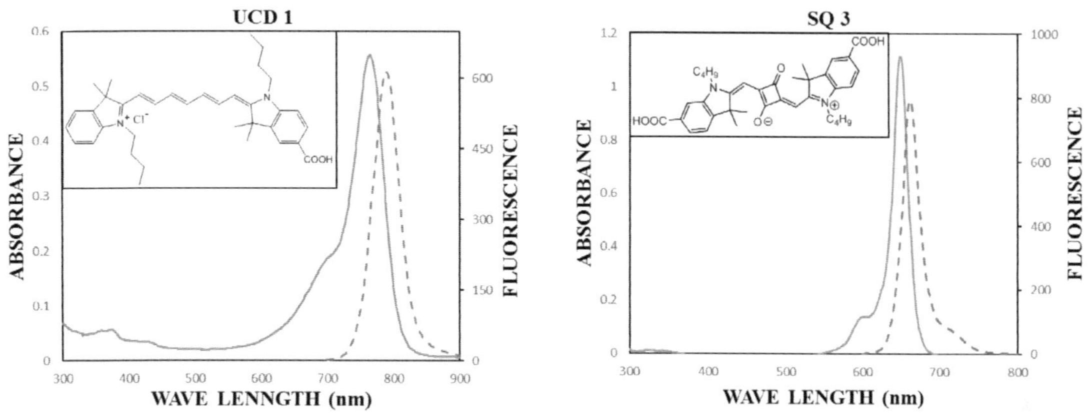

Figure. 1 Electronic absorption (solid line) and fluorescence emission spectra (dashed line) of UCD-1 (left) and SQ-3 (right) in DMF solution (5μM). Molecular structures of dyes are shown in the inset.

Table 1. Spectral properties of sensitizing dyes in DMF and 0.1 M PBS solutions at pH 7.4.

DYE	DMF solution		PBS solution		Stoke Shift		(ε)
	λ (max) Absorption	λ (max) Emission	λ (max) Absorption	λ (max) Emission	DMF	PBS	($dm^3 M^{-1} cm^{-1}$)
UCD -1	764 nm	788 nm	750 nm	770 nm	24 nm	20 nm	1×10^5
SQ - 3	650 nm	662 nm	639 nm	644 nm	12 nm	5 nm	2×10^5

Phosphate buffer solution (PBS) has been most commonly employed for biological studies especially in order to investigate the interaction between dyes and biomolecules for imaging applications. Keeping this in mind absorption and emission spectra of these dyes were also recorded in the 0.1 M PBS solution at pH 7.4 and shown in the Fig. 2. It is worth mentioning that in the case of phosphate buffer solution (PBS) both cyanine and squaraine dyes although exhibit nearly similar absorption and emission spectral behavior but slightly blue-shifted as compared to that observed in the DMF solution. This blue shift of λ_{max} for both of the dyes in PBS could be attributed to enhanced dye aggregation promoted by the hydrogen bonding between the dye molecules due to the presence of –COOH functional groups. It is well known that both cyanine and squaraine exhibit dye aggregation solution owing to their flat molecular structure [27]. Another interesting feature is that fluorescence emission intensity of cyanine dye is highly decreased (about 10 times) as compared to squaraine dyes for similar concentration and experimental measurement conditions. This suggests that cyanine dyes could be a good candidate as quenching probe while squaraine dyes as fluorescence probe for the design and development of NIR fluorescence resonance energy transfer systems.

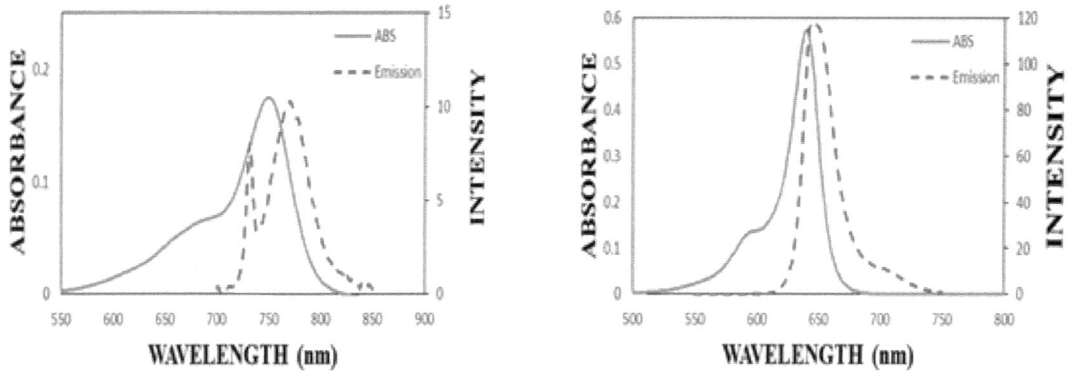

Figure. 2 Electronic absorption and fluorescence emission spectra for 2μM concentration of UCD-1 (left) and SQ-3 (right) in 0.1 M PBS solution.

3.2. Interaction of dyes with BSA

In order to avoid chemical reactions and further purification steps of chemical labelling of biomolecules with dyes in optical bio-imaging, non-covalent labelling is associated with investigation of dye-protein interaction are getting increasing attention. Williams *et al* are amongst one of the pioneer researchers to study about the non-covalent interactions between the NIR dyes and human serum albumin (HSA) [28]. Bovine serum albumin (BSA) is one of the most widely investigated model protein to investigate drug-protein interactions owing to its high sequence identity to that of human serum albumin [29]. Keeping this in mind, interaction of BSA as a model protein has been executed using different amounts of protein. Figure 3 exhibits the electronic absorption spectra of PBS buffer solution of dyes UCD-1 and SQ-3 in the presence and absence of BSA. The spectra show two different sets of electronic absorption transitions associated with absorption regions of the protein BSA (250 to 300 nm) and squaraine/cyanine dyes (650 to 800 nm). The absorption region of squaraine dye is confined in 550 to 700 nm while cyanine dyes absorbs strongly in the 600-800 nm.

It can be observed that increasing BSA concentration leads to gradual increase in the intensity of absorption between 250 to 300 nm which associated with protein absorption. However, the dye absorption region exhibited differential behaviour with UCD-1 and SQ-3. The BSA interaction with cyanine dye (UCD-1) exhibits the random increase and decrease of absorption intensity corresponding to the dye absorption around 750 nm and slight red-shift from 758 nm to 768 nm upon the addition of BSA. Slight red-shift indicates the suppression of dye aggregation by BSA molecules and indicates the presence of interaction between dye and BSA molecules. At the same time, random increase and decrease without any isobestic point indicates that interaction between BSA and UCD-1 is not so simple and a detailed investigation on site specific binding is necessary [14]. A similar kind of complicated BSA-cyanine dye interaction with random increase and decrease absorption intensity at λ_{max} along with red-shift has also been reported by Pisoni *et al* [23].

On the other hand, interaction of BSA with SQ-3 exhibits interesting behaviour apart from slight red-shift from 639 nm to 649 nm upon increasing concentration of BSA. At the fix dye concentration of 2μM, there is an initial decrease in absorption intensity (hypochromism) corresponding to monomeric SQ-3 absorption around 4-5μM along with clear isobestic point and then remains constant upon the further addition of BSA. This red shift and hypochromicity observed for interaction between SQ-3 and BSA could be attributed to the presence of hydrophobic environment provided by BSA to the hydrophilic SQ-3 and is indicative for the formation of squaraine dye-BSA conjugate. Similar electronic absorption behaviour with clear isobestic point and hypochromism has also been noticed by Jisha et al during investigation of interaction between squaraine dyes with HSA [30].

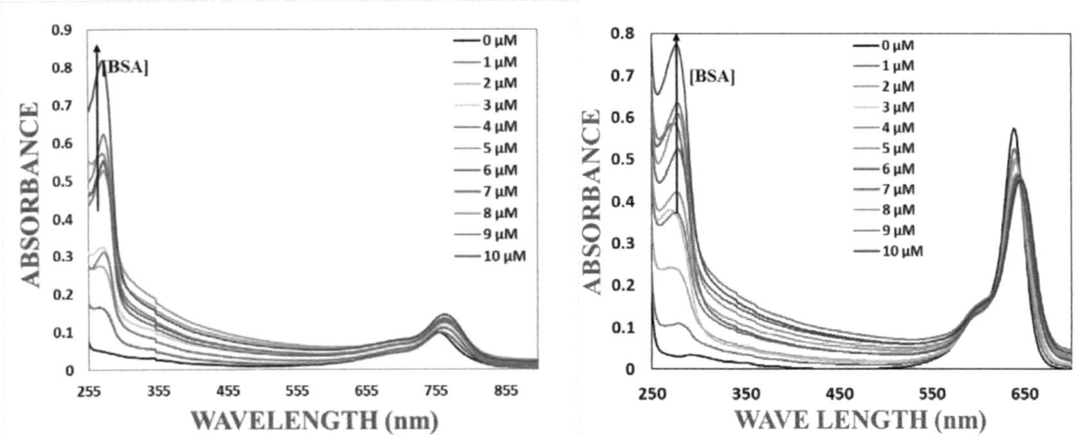

Figure. 3 Electronic absorption of UCD-1 (left) and SQ-3 (right) in 0.1 M PBS at different BSA concentrations for a fixed dye concentration of 2μM.

Changes in the fluorescence behavior of dyes as a result their interactions with biomolecules like proteins, DNA etc. forms fundamental basis for fluorescence bio-imaging. Spectroscopic behaviour of fluorophoric cyanine and squaraine dyes have been reported to be highly sensitive to the environmental conditions owing to their self-aggregation or interaction with other existing molecules in their vicinity [31]. In general fluorophores bind with protein covalently or non-covalently and non-covalent binding finds its application in biomedical science in order to investigate the changes in the conformation of proteins and their binding with drugs [32]. Fluorescence emission spectra of both of the dyes UCD-1 and SQ-3 as function of BSA concentration is shown in the Fig. 4. It can be clearly seen that both of the dyes exhibit increase in the fluorescence intensities of dyes near their respective peak maxima along with the red-shit upon addition of increasing concentration of BSA at fix dye concentration. This increase in the fluorescence intensity along with the red-shift in the peak maxima suggests the interaction between the dye molecules and BSA due to non-covalent BSA-dye conjugate formation. It has been reported that site specific binding of BSA using Dansylproline (DP) and Dansylamide (DNSA) as site specific binder and squaraine dyes as fluorophore leads to decrease in the fluorescence intensity of BSA-dye conjugate as a function of site specific binder concentration [33].

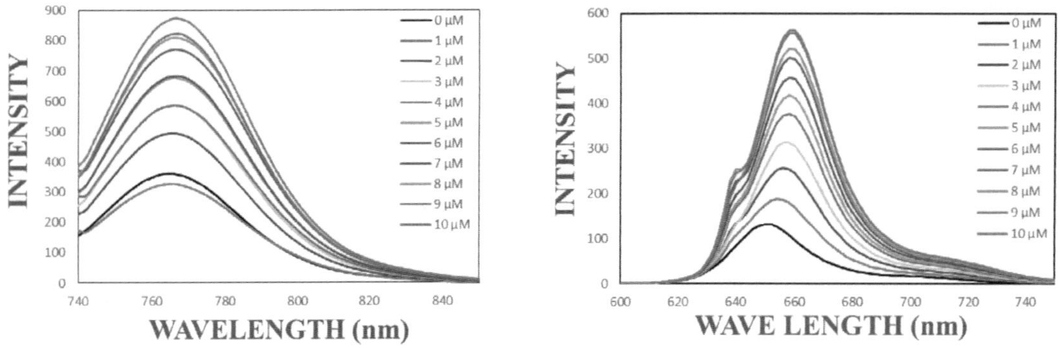

Figure. 4 Fluorescence emission spectra of UCD-1 (left) and SQ-3 (right) in 0.1 M PBS with varied BSA concentrations for a fixed dye concentration of 2μM.

This suggests that in present case also enhancement in fluorescence intensity in the presence of BSA for both of the dyes UCD-1 and SQ-3 is due to non-covalent binding with BSA. To compare the binding ability and relative association of these two dyes with BSA, apparent binding constant (K_a) was estimated using equation (2) mentioned in the experimental section and shown in the Fig. 5. From equation 2, it can be seen that slope gives the value K_a which was estimated to be 7.08 x 10^6 M^{-1} and 6.22 x 10^6 M^{-1} for dyes UCD-1 and SQ-3, respectively. For non-covalent binding with BSA using squaraine dyes, K_a in the similar range has also been reported [30].

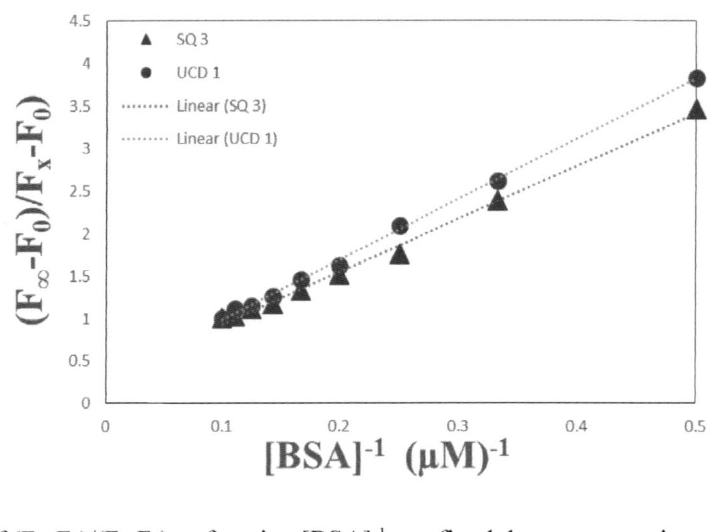

Figure. 5 Plot of $(F_\infty-F_0)/(F_x-F_0)$ as function $[BSA]^{-1}$ at a fixed dye concentration of 2.0×10^{-6} M.

At the same time cyanine dye (UCD-1) not only exhibits the higher binding affinity as compared to squaraine dye (SQ-3) but also an order of higher magnitude K_a had been reported for typical NIR cyanine dyes having non-functionalized hydrophilic groups [23]. In the present case, this could be associated with presence of direct ring substituted –COOH group promoting hydrogen bonding with binding sites of BSA. In this situation, one can argue that why SQ-3 having two such -COOH functionality shows relatively lower binding affinity as compared to UCD-1. Exact reason is although not completely clear but it might be associated judicious balance of main hydrophobic core and presence of active group for promoted binding with the active sites. Actually, active site of model protein BSA used in this work is covered by hydrophobic amino acid residues to form the binding sites thus hydrophobic region make the outer core for interacting with dye molecules.

4. Conclusions

Direct ring carboxy functionalized far-red to NIR sensitive dyes SQ-3 and UCD-1 has been successfully synthesized and characterized in order to explore their applicability as fluorescent probes. Interaction of these dyes with BSA as model protein leads to enhancement in the fluorescence intensity along with bathochromic shift in the emission maxima suggesting the dye-BSA conjugate formation. Both the dyes show very high affinity for their association with BSA. Site specific binding is still needed to have some more insight about preference of these dyes for the particular binding sites of the BSA.

Acknowledgments
Authors would like to acknowledge the support from Prof. Shigeori Takenaka (Kyushu Institute of Technology, Tobata) in obtaining Mass spectra.

References

[1] Weissleder R and Pittet M J 2008 *Nature*. **452** 580-589.
[2] Kobayashi H, Ogawa M, Alford R, Choyke P L and UranoY 2009 *Chem. Rev.* **110** 2620-2640.
[3] Law K Y 1992 *Chem. Mater*. **4** 605-611.
[4] Chen H, Farahat M S, Law K Y and Whitten D G 1996 *J. Am. Chem. Soc.* **118** 2584-2594.
[5] Ahn H Y, Yao S, Wang X and Belfield K D 2012 *ACS Appl. Mater. Interfaces*. **4** 2847-2854.
[6] McEwen J J and Wallace K J 2009 *Chem. Commun.* 6339-6351.
[7] Johnson J R, Fu N, Arunkumar E, Leevy W M, Gammon S T, Piwnica-Worms D and Smith B D 2007 *Angew. Chem*. **119** 5624-5627.
[8] Gassensmith J J, Arunkumar E, Barr L, Baumes J M, DiVittorio K M, Johnson J R, Noll B C and Smith B D 2007 *J. Am. Chem. Soc.* **129** 15054-15059.
[9] Yi X, Wang F, Qin W, Yang X and Yuan J 2014 *Int J Nanomedicine*. **9** 1347.
[10] Mishra A, Behera R K, Behera P K, Mishra b K and Behera G B 2000 *Chem. Rev.* **100** 1973-2012.
[11] Yuan L, Lin W, Zheng K, He L and Huang W 2013 *Chem. Soc. Rev.* **42** 622-661.
[12] El-Shishtawy R M 2009 *Int. J. Photoenergy*.
[13] Benzi C, Bertolino C A, Miletto I, Ponzio P, Barolo C, Viscardi G, Coluccia S and Caputo G 2009 *DYES PIGMENTS*. **83** 111-120.
[14] Tatikolov A S and Costa S I M 2004 *Biophysical chemistry*. **107** 33-49.
[15] Reddington M V 2007 *Bioconjugate Chem.* **18** 2178-2190.
[16] Gonçalves M S T 2008 *Chem. Rev.* **109** 190-212.
[17] Oswald B, Lehmann F, Simon L, Terpetschnig E and Wolfbeis O S 2000 *Anal. Biochem* **280** 272-277.
[18] Pansare V J, Hejazi S, Faenza W J and Prud'homme R K 2012 *Chem. Mater*. **24** 812-827.
[19] Povrozin Y A, Kolosova O S, Obukhova O M, Tatarets A L, Sidorov V I, Terpetschnig E A and Patsenker L D 2009 *Bioconjugate Chem*. **20** 1807-1812.
[20] Zhang Y Z, Yang Q F, Du H Y, Tang Y L, Xu G Z and Yan W P 2008 *Chin. J. Chem.* **26** 397-401.
[21] Inoue T, Pandey S S, Fujikawa N, Yamaguchi Y and Hayase S 2010 *J. Photochem. Photobiol. A*. **213** 23-29.
[22] Pandey S S, Inoue T, Fujikawa N, Yamaguchi Y and Hayase S 2010 *J. Photochem. Photobiol. A*. **214** 269-275.
[23] Pisoni O S, Todeschini L, Borges A C s A, Petzhold C L, Rodembusch F S and Campo L F

Journal of Physics: Conference Series **704** (2016) 012012 doi:10.1088/1742-6596/704/1/012012

2014 *J. Org. Chem.* **79** 5511-5520.
[24] Berezin M Y, Guo K, Teng B, Edwards W B, Anderson C J, Vasalatiy O, Gandjbakhche A, Griffiths G L and Achilefu S 2009 *J. Am. Chem. Soc.* **131** 9198-9200.
[25] Redy O, Kisin-Finfer E, Sella E and Shabat D 2012 *Org. Biomol. Chem.* **10** 710-715.
[26] Benesi H A and Hildebrand J 1949 *J. Am. Chem. Soc.* **71** 2703-2707.
[27] Tatikolov A S 2012 J. *Photochem. Photobiol, C.* **13** 55-90.
[28] Williams R J, Lipowska M, Patonay G and Strekowski L 1993 *Anal. Chem.* **65** 601-605.
[29] Alarcón E, Edwards A M, Garcia A M, Munoz M, Aspée A, Borsarelli C D and Lissi E A 2009 *Photochem. Photobiol. Sci.* **8** 255-263.
[30] Jisha V S, Arun K T, Hariharan M and Ramaiah D 2010 *J. Phys. Chem. B.* **114** 5912-5919.
[31] Arun K T and Ramaiah D 2005 *J. Phys. Chem. A.* **109** 5571-5578.
[32] Patonay G, Salon J, Sowell J and Strekowski L 2004 *Molecules.* **9** 40-49.
[33] Zhang Y, Yue X, Kim B, Yao S, Bondar M V and Belfield K D 2013 *ACS Appl. Mater. Interfaces.* **5** 8710-8717.

Journal of Physics: Conference Series **704** (2016) 012013 doi:10.1088/1742-6596/704/1/012013

Functionalized vertically aligned ZnO nanorods for application in electrolyte-insulator-semiconductor based pH sensors and label-free immuno-sensors

Narendra Kumar[1,3], Sujata Senapati[1], Satyendra Kumar[2,3], Jitendra Kumar[1], and Siddhartha Panda[1,2,3,*]

[1]Materials Science Programme, Indian Institute of Technology Kanpur, Kanpur-208016, India

[2]Department of Chemical Engineering, Indian Institute of Technology Kanpur, Kanpur-208016, India

[3]Samtel Centre for Display Technologies, Indian Institute of Technology Kanpur, Kanpur-208016, India

*E-mail:- spanda@iitk.ac.in

Abstract. Vertically aligned ZnO nanorods were grown on a SiO_2/Si surface by optimization of the temperature and atmosphere for annealing of the seed. The seed layer annealed at 500 °C in vacuum provided well separated and uniform seeds which also provided the best condition to get densely packed, uniformly distributed, and vertically aligned nanorods. These nanorods grown on the substrates were used to fabricate electrolyte-insulator-semiconductor (EIS) devices for pH sensing. Etching of ZnO at acidic pH prevents the direct use of nanorods for pH sensing. Therefore, the nanorods functionalised with 3-aminopropyltriethoxysilane (APTES) were utilized for pH sensing and showed the pH sensitivity of 50.1 mV/pH. APTES is also known to be used as a linker to immobilize biomolecules (such as antibodies). The EIS device with APTES functionalized nanorods was used for the label free detection of prostate-specific antigen (PSA). Finally, voltage shifts of 23 mV and 35 mV were observed with PSA concentrations of 1 ng/ml and 100 ng/ml, respectively.

1. Introduction

ZnO nanostructures have been extensively studied to develop various potentiometric and field effect based chemical and bio-sensors because of their large surface area to volume ratio and compatibility with biomolecules [1-9]. The chemical functionalization of these nanostructures with 3-aminopropyltriethoxysilane (APTES) prevents etching at acidic pH values [10], and also helps to improve their pH sensing characteristics as obtained in a functionalized ZnO nanowires based liquid gated transistor [11]. The amine groups of the APTES and the polyethylenimine (PEI) are also used as linker molecules to immobilize capture biomolecules (proteins) to develop different immunoassays [12]. Therefore, in addition to their protection at acidic pH values, the improved pH sensitivity of chemically functionalized ZnO nanorods (NRs) could also obviate the need of chemical functionalization and high-k dielectrics such as Si_3N_4, Al_2O_3, and Ta_2O_5 in the fabrication of the enzymatic bio-sensors [13-15]. In order to maximize the surface area to volume ratio of sensing surface, it is necessary to obtain vertically aligned ZnO nanorods. The microstructure of the seed layer is crucial to obtain the desired orientation and density of the nanorods which could be controlled by the concentration of the ZnO colloid and the annealing temperature of the deposited films [16, 17]. Therefore, it is necessary to obtain a seed layer with well separated and uniform seed particles by varying its annealing temperature and environment. Recently, an electrolyte-insulator-semiconductor (EIS) structure using RF sputtered ZnO thin film was used for pH and multiple ion detections [18]. As per the best of our knowledge, we are not aware of any report on the utilization of functionalized nanorods in the EIS structure for pH and label free detection of prostate-specific antigen (PSA).

In this work, first the effect of temperature and atmosphere on annealing of the ZnO seed layer was studied to obtain the densely packed and vertically aligned nanorods. The properties of grown nanorods were studied with an X-ray diffraction and UV-Vis absorption. Then the as grown and functionalized nanorods grown over SiO_2/Si substrates were studied with field emission scanning electron microscope (FESEM) and X-ray photoelectron spectroscopy (XPS) analysis. The pH sensing characteristics of the EIS devices with the bare and the functionalized nanorods were studied followed by their application for label free detection of PSA.

2. Fabrication and characterization

2.1. Seed layer preparation, growth of nanorods, and chemical functionalization

The 0.75 M zinc acetate dihydrate (ZAD) was dissolved in 40 ml isopropyl alcohol (IPA) followed by stirring the mixture at 60 ^{0}C for 45 min at 1000 rpm in a 50 ml glass autoclavable bottles caps, stirred at 1000 rpm for 15 min at room temperature. Subsequently, the complexing agent monoethanolamine (MEA) in equimolar of zinc ions was added at 60 ^{0}C and stirred for another 15 min at 1000 rpm till solution became transparent. To get the fine seed solution, the solution was heated in two steps- in a vacuum oven first for 1 hour at 120 ^{0}C, and then at 80 ^{0}C for 2 hour, and subsequently cooled down to room temperature. The prepared seed solution was spin coated on a piranha cleaned SiO_2/Si substrates at 400 rpm for 1minute and post baked for 10 minutes at 250 °C on a hot plate. To further improve the seed quality, the films were further annealed at different temperatures in air and vacuum. To grow the nanorods, aqueous solutions of 2 mM zinc nitrate hexahydrate (ZHN) with equimolar hexamine was prepared in 50 ml deionized (DI) water by stirring at 60 ^{0}C for 1 hr. The prepared ZnO seed layer over SiO_2/Si substrates were dipped in the prepared solution and the vessel was sealed, followed by heating for 4h at 90°C. After growth of nanorods the samples were thoroughly washed with a stream of DI water to remove the adsorbed nanorods and dried with nitrogen. APTES of purity 99.9% was purchased from Sigma Aldrich, Germany. For the functionalization with APTES, nanorods grown samples were kept in 3% APTES in toluene for 24h and washed with warm toluene, ethanol, and dried with nitrogen.

2.2. EIS device fabrication and characterization

The process steps to fabricate the EIS device were similar to those used in our previous works [19, 20]. The SiO_2/Si wafers of size 1.5 x 1.5 cm^2 with grown ZnO nanorods were processed using photolithography of SU-8 epoxy for the fabrication of a double step reservoir. The front and back contacts were fabricated by depositing Au (10 mm x 0.3 mm) and Al over the second step of the reservoir and back side of SiO_2 etched silicon, respectively. The Ag/AgCl ink was screen printed over the gold electrode using a mask of 6 mm x 0.5 mm to make the integrated reference electrode. Capacitance-voltage measurement (C-V) measurements were conducted with the help of an impedance analyzer (Model: Agilent 4294A). The sensitivity was measured by normalizing the C-V curves and measuring the reference voltage at $0.5C_{normalized}$ with buffer solutions of the respective pH values followed by linear fitting of the reference voltages versus the pH values [19]. The EIS device with APTES functionalized nanorods was treated with 10% glutaraldehyde in DI water for 1h, washed with DI water and dried in nitrogen. Then, the antibodies (anti-PSA) were attached by filling the reservoir with 20 µl of anti-PSA solution of concentration 0.1 mg/ml in phosphate buffer saline (PBS) for 16h at 4 °C and washed thrice with PBS, and dried in nitrogen. The PSA antigens were bound by filling the reservoir of 20 µl PSA solution of different concentrations range of 1 ng/ml to 1 µg/ml for 30 minutes at 37 °C. FESEM (Model: JSM-7100F; JEOL) for the imaging of ZnO seed and nanorods. X-ray diffractometer (Model: Panalytical XPert) and spectrophotometer (Model: Varian 5000) were used for phase analysis and band gap measurement, respectively.XPS (PHI 5000 Versa Probe II) was used for the surface analysis of the as grown and functionalized nanorods.

3. Results and discussion

3.1. Effect of seed layer annealing

Growth of vertically aligned ZnO nanorods was realized with a hydrothermal process by optimization of seed layer. The prepared seed layers were annealed at 400 and 500 °C in air as well as in vacuum. As can be seen from figure 1, seed layer annealed in vacuum showed an increase in the average seed size from 16.6±1.4 nm to 25.8±3.8 nm with increase in the annealing temperature from 400 to 500 °C. While no significant changes were observed when annealed in air except flaky structures were observed upon annealing at 500 °C. No growth of nanorods was seen on as prepared seed layer (figure1b) while few adsorbed microrods grown in solution remained on the surface. Some smaller nanorods were grown on the substrate with seed layers annealed at 400 °C in air as well as vacuum. No growth of nanorods was observed on the seed layer annealed at 500 °C in air because of the flaky nature of the seed structure. The best (vertical) growth of nanorods (figure 1j) was obtained on the seed layer annealed at 500 °C in vacuum. Therefore, one can conclude that the seed layer with well separated particles facilitates to obtain uniform and vertically aligned nanorods.

Figure 1. Seed layer as prepared (a) bare (b) with nanorods, Seed layer annealed at 400 °C in air (c) bare (d) with nanorods, Seed layer annealed at 400 °C in vacuum (e) bare (f) with nanorods, Seed layer annealed at 500 °C in air (g) bare (h) with nanorods, Seed layer annealed at 500 °C in vacuum (i) bare (j) with nanorods.

The nanorods grown on the substrate with the seed layer annealed at 500 °C in vacuum were characterized with XRD and UV-Vis absorption and utilized in the fabrication of the EIS device. The XRD pattern shown in figure 2 confirmed the hexagonal wurtzite structure of zinc oxide nanorods with the prominent peaks corresponding to (100), (002), (101), and (103) planes with three smaller peaks corresponding to planes (102), (110), and (112) [21]. The UV-Vis absorption spectra showed the absorption band at wavelength of 383 nm which corresponds to the band gap of 3.24 eV comparable to the reported literature value of 3.21 eV [21].

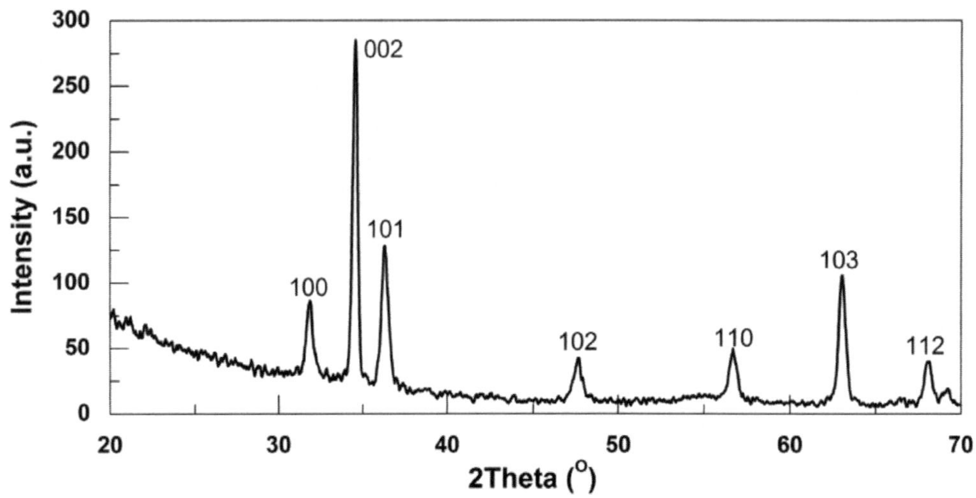

Figure 2. XRD pattern of ZnO NRs grown over SiO_2/Si substrate with the seed layer annealed at 500 °C in vacuum. (The peak indices for hexagonal system are in 3-index notation)

3.2. Effect of exposure at acidic pH

The fabricated EIS devices with bare ZnO nanorods were tested for pH detection with pH values decreasing from 9.2 to 4.3. As shown in figure 4, little variation in the capacitance was observed with no significant change in flatband voltage between pH 9.2 to 7.3. Increased capacitance upon exposure of the device with pH 4.3, as seen in figure 3a,could be attributed to the additional layer of ZnO seed and nanorods being etched, and this was supported by the inspection of micrographs, one such being shown in figure 3b, taken after exposing the bare ZnO nanorods based samples with buffer solution of pH 4.3. Also, this was further confirmed as the device showed C-V characteristics and pH sensitivity usually seen for the device only with the SiO_2 dielectric. There are reports available where RF sputtered ZnO films over silicon annealed above 600 °C, A-ZnO nanostructures over glass, ZnO nanowires over Cr/Au substrates were utilized for pH sensing [6, 18, 22]. Thus, the etching of ZnO seed and nanorods at acidic pH could be attributed to the poor interface at ZnO/SiO_2. In order to overcome this problem, grown nanorods were functionalized with APTES and tested with solution of pH 4.3 and no etching of the surface was observed (micrographs not shown here).

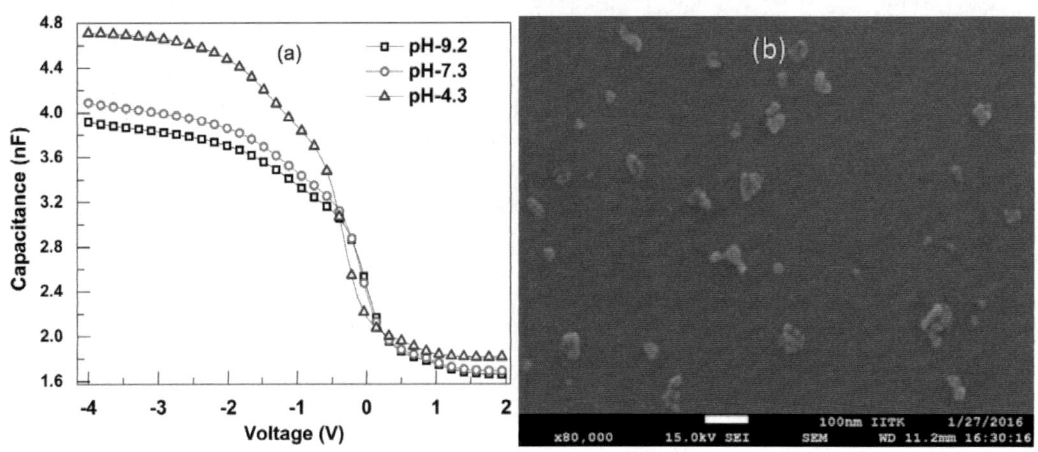

Figure 3. (a) C-V curves of EIS device with bare ZnO nanorods at varying pH values, (b) The micrograph of the sample exposed to pH 4.3.

3.3. Effect of APTES functionalization

The micrographs of ZnO NRs before and after functionalization are shown in figure 4a and 4b, respectively. Figure 4b shows uniformly coated nanorods with APTES resulting in the blunt edges with slightly increased diameter [10].Figure 5a shows the XPS spectra of bare (as grown) ZnO NRs with three peaks corresponding to C 1s (carbon), O1s (oxygen), and Zn 2p3 (zinc) at 283.6 eV, 529.4 eV, and 120.3 eV, respectively [23]. Figure 5b shows the XPS spectra with two additional peaks of N 1s (nitrogen) at 397.9 eV and Si 2p (silicon) at100.8 eV in APTES functionalized nanorods and these are attributed to the amine groups (-NH_2) and silane, respectively, in the APTES structure [23-25]. Moreover, the atom weight % of C 1s increased from 16.1 to 29.9 because of ethylene and ethoxy groups in APTES while it significantly decreased from 52.9 to 35.9 for O1s and from 31.0 to 3.0 for Zn 2p3 upon functionalization due to the coverage of ZnO surface with the APTES molecules.

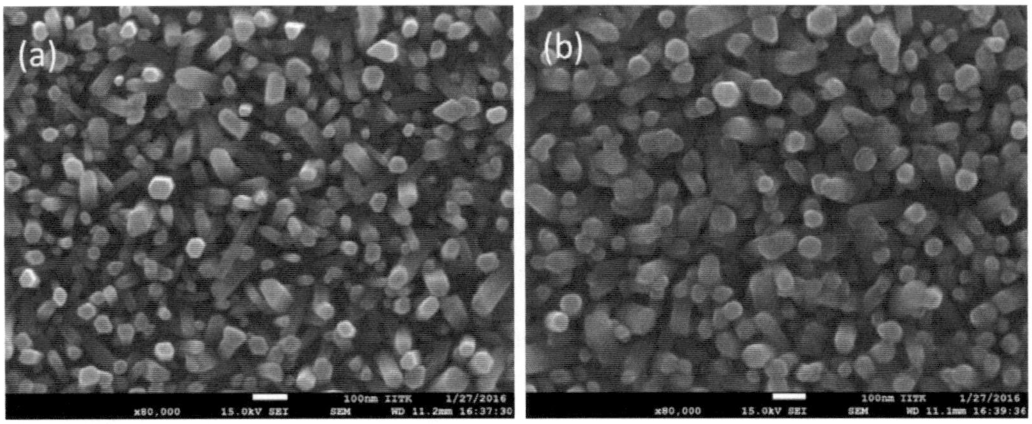

Figure 4. (a) As grown ZnO NRs, (b) APTES functionalized ZnO NRs.

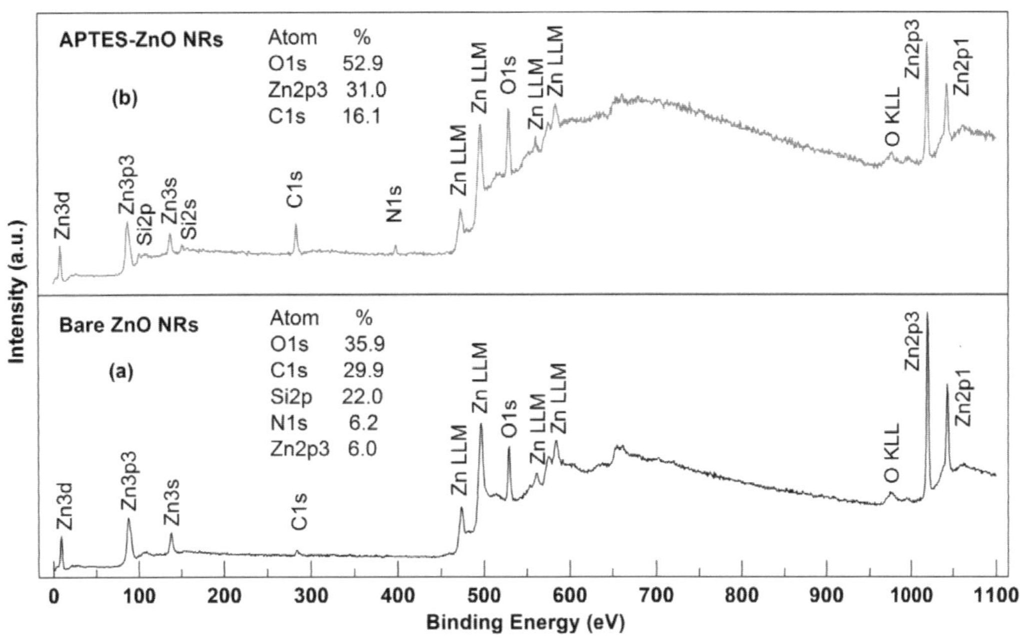

Figure 5. XPS spectra of (a) as grown ZnO NRs (b) APTES functionalized ZnO NRs.

3.4. pH sensing characteristics of functionalized nanorods

Figure 6a shows the C-V curves of the EIS device with APTES functionalized nanorods at varying pH values from 4.3 to 9.2. C-V curves shift towards the positive with increase in pH values because of the increased negative surface charge at the electrolyte/ZnO interface with increasing pH values. Increased negative surface charges induce accumulation in the p-Si without any applied bias resulting in a positive flatband voltage shift. The measured reference voltages at $0.5C_{normalized}$ with the varying pH values were plotted with respect to pH and the slope obtained by linearly fitting of the plot gives the pH sensitivity. The EIS device with APTES functionalized nanorods showed the average sensitivity of 50.1 mV/pH. The pH sensitivity obtained in this work is higher than the recently reported value (43.2 mV/pH) of APTES functionalized nanowires used in extended gate field effect transistors [22]. Yang et al. [6] obtained a maximum sensitivity of 57.9 mV/pH using Al doped ZnO nanostructures in an extended gate field effect transistor using commercial MOSFETs with high-K dielectrics. However in this work, the SiO_2 was used as dielectric materials to fabricate the EIS device which has a relatively lower dielectric constant (3.9). Therefore, the sensitivity of our devices could further improved by utilizing high-K dielectric materials.

APTES modified ZnO surface contains both type of surface groups –ZnOH and $-NH_2$, as confirmed from the XPS spectra of O 1s and N 1s of functionalized nanorods. A similar surface nature (–OH and $-NH_2$ groups) has been reported on APTES modified silicon and ZnS/Silica nanowires, which showed a linear pH response [26, 27]. The $-NH_2$ groups get protonated to $-NH_3^+$ at acidic pH values while –ZnOH groups get deprotonated to ZnO at alkaline pH values as reported in our previous work for a SiO_2 surface textured containing –SiOH and $-NH_2$ groups with silica nanoparticles via APTES as the linker [28]. Therefore, in addition to the protection of ZnO nanorods from etching at acidic pH values, the APTES modification resulted better the pH sensitivity with a better linearity of 0.99.

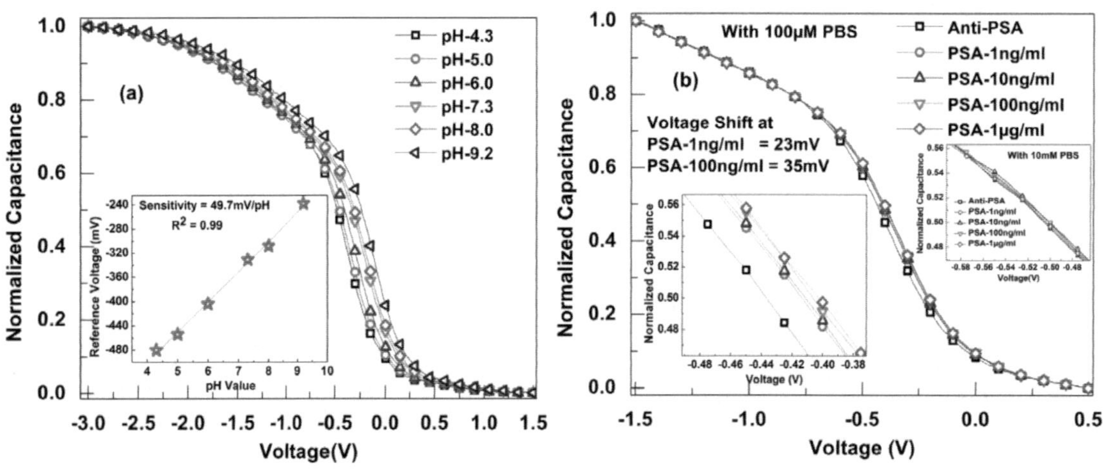

Figure 6. C-V curves of EIS device with APTES functionalized ZnO NRs (a) at varying pH values (the pH sensitivity plot in the inset) (b) at varying PSA concentration measured at 100 μM PBS (for 10mM PBS in the right inset).

3.5. Label-free PSA detection

As discussed in the introduction, in addition to the enhancement of sensitivity of pH detection, the APTES (which acts as a linker) functionalized ZnO nanorods allow the covalent immobilization of biomolecules to utilize them in the development of immuno-sensors. The label-free detection of PSA biomarker has been realized using the EIS device with APTES functionalized nanorods. The PSA biomarker possesses a net negative surface charge at the physiological pH of 7.4. Therefore, the PSA (antigen) binding with immobilized anti-PSA (antibody) on the sensing surface of EIS device induces the accumulation of holes (majority carriers) in the semiconductor (p-silicon) side. Thus the accumulation of holes without any applied bias results in a positive shift in the flatband voltage. To perform the PSA detection experiment, two strengths of PBS solutions i.e. 10 mM and 100 μM with 25 mM KCl as ionic strength adjuster [29] were used by taking consideration of the Debye screening effect which remarkably reduces the sensitivity of the device in label-free detection of bioanalytes. As shown in figure 6b (inset right), no significant voltage shift in the C-V curve observed upon binding of PSA of concentration 1ng/ml with 10 mM PBS. While about 23 mV (figure 6b) voltage shift obtained using the PBS solution of 100 μM at pH 7.4. This notable voltage shift after antigen binding is attributed to the increased Debye screening length with decreased PBS strength [30]. The voltage shift was further increased to a maximum of 35 mV with increased PSA concentration of 100 ng/ml without removing the previously bonded antigens at 1 ng/ml. As shown in figure 6b (left inset), no further voltage shift was observed at higher PSA concentration of 1 μg/ml, which indicates the maximum numbers of available antibodies immobilized over sensor surface were engaged with antigens of concentration till 100 ng/ml. The calibration curve can be obtained by regenerating the immobilized surface with antibodies by removing the antigen after detection at each concentration and this is the scope of future work.

4. Conclusion

To grow vertically aligned ZnO nanorods with uniform distribution over SiO_2/Si substrates, the seed layer annealing temperature was optimized by annealing in air and vacuum to obtain well separated and uniform seed size. The seeds were not distinguishable in seed layer post baked at 250 °C and formation of nanorods was not observed. Then the samples were annealed at 400 °C and 500 °C for 1h in air as well as in vacuum. The seed layer annealed at 500 °C in vacuum provided well separated and uniform seeds which also proven to be the best condition to get densely packed, uniformly distributed, and vertically aligned nanorods. These SiO_2/Si substrates with nanorods were subsequently utilized to develop an EIS device for pH sensing. The etching of ZnO at an acidic pH 4.3 restricted their direct use for pH sensing. The nanorods were then functionalised with 3-aminopropyltriethoxysilane (APTES) and utilized for pH sensing which showed a pH sensitivity of 50.1 mV/pH. The APTES is also known to be used as linkers to immobilize the biomolecules (antibodies). Thus, the EIS device with APTES functionalized nanorods was used for the label free detection of prostate-specific antigen. Finally, voltage shifts of 23 mV and 35 mV were observed with the PSA concentrations of 1 ng/ml and 100 ng/ml, respectively.

Acknowledgment

The financial support of the DST Science and Engineering Research Board, India (Grant number SB/S3/CE/055/2013) is acknowledged.

References

[1] Al-Hilli S M, Willander M, Öst A and Strålfors P 2007 *J. Appl. Phys.* 102 084304
[2] Ogata K-I, Koike K, Sasa S, Inoue M and Yano M 2009 *J. Vac. Sci. Technol. B* 27 1684-7
[3] Kang B S, Wang H T, Ren F, Pearton S J, Morey T E, Dennis D M, Johnson J W, Rajagopal P, Roberts J C, Piner E L and Linthicum K J 2007 *Appl. Phys. Lett.* 91 252103
[4] Fulati A, Usman Ali S, Riaz M, Amin G, Nur O and Willander M 2009 *Sensors* 9 8911
[5] Usman Ali S M, Nur O, Willander M and Danielsson B 2009 *IEEE Trans. Nanotech.* 8 678-83
[6] Yang P-Y, Jyh-Liang W, Chiu P-C, Jung-Chuan C, Cheng-Wei C, Hung-Hsien L and Huang-Chung C 2011 *IEEE Elec. Dev. Lett.*, 32 1603-5
[7] Ahmad R, Tripathy N, Jung D-U-J and Hahn Y-B 2014 *Chem. Comm.* 50 1890-3
[8] Bhattacharya A, Rao V P, Jain C, Ghose A and Banerjee S 2014 *Mater. Lett.* 117 128-30
[9] Ibupoto Z H, Mitrou N, Nikoleli G-P, Nikolelis D P, Willander M and Psaroudakis N 2014 *Electroanalysis* 26 292-8

[10] Loh L, Briscoe J and Dunn S 2015 *ACS Appl. Mater. Interfac.* 7 152-7
[11] Pachauri V, Vlandas A, Kern K and Balasubramanian K 2010 *Small* 6 589-94
[12] Yakovleva J, Davidsson R, Lobanova A, Bengtsson M, Eremin S, Laurell T and Emnéus J 2002 *Anal. Chem.* 74 2994-3004
[13] Dzyadevych S V, Soldatkin A P, El'skaya A V, Martelet C and Jaffrezic-Renault N 2006 *Anal. Chim. Act.* 568 248-58
[14] Vijayalakshmi A, Tarunashree Y, Baruwati B, Manorama S V, Narayana B L, Johnson R E C and Rao N M 2008 *Biosens. Bioelectron.* 23 1708-14
[15] Stepurska K V, Soldatkin O O, Arkhypova V M, Soldatkin A P, Lagarde F, Jaffrezic-Renault N and Dzyadevych S V 2015 *Talanta* 144 1079-84
[16] Teng M, Min G, Mei Z, Yanjun Z and Xidong W 2007 *Nanotechnology* 18 035605
[17] Tao Y, Fu M, Zhao A, He D and Wang Y 2010 *J. Alloy. Comp.* 489 99-102
[18] Haur Kao C, Chen H, Ling Lee M, Chun Liu C, Ueng H-Y, Cheng Chu Y, Jie Chen Y and Ming Chang K 2014 *J. App. Phys.* 115 184701
[19] Kumar N, Kumar J and Panda S 2015 *AIP Adv.* 5 067123
[20] Kumar N, Tiwari A P, Kumar J and Panda S 2015 *2nd Inter. Symp. Phys. Tech. of Sens. (ISPTS)*,Pune, India: IEEE pp 214-8
[21] Foo K L, Hashim U, Muhammad K and Voon C H 2014 *Nanos. Res. Lett.* 9 429
[22] Qing Z, Wenpeng L, Chongling S, Hao Z, Wei P, Daihua Z and Xuexin D 2015 *Nanotechnology* 26 355202
[23] Lee J, Choi S, Bae S J, Yoon S M, Choi J S and Yoon M 2013 *Nanoscale* 5 10275-82
[24] Haddada M B, Blanchard J, Casale S, Krafft J-M, Vallée A, Méthivier C and Boujday S 2013 *Gold Bullet.* 46 335–41
[25] Zhang W, Luo X-J, Niu L-N, Yang H-Y, Yiu C K Y, Wang T-D, Zhou L-Q, Mao J, Huang C, Pashley D H and Tay F R 2015 Carriers *Sci. Rep.* 5 11199
[26] Cui Y, Wei Q, Park H and Lieber C M 2001 *Science* 293 1289-92
[27] He J H, Zhang Y Y, Liu J, Moore D, Bao G and Wang Z L 2007 *The J. Phys. Chem. C* 111 12152-6
[28] Kumar N, Kumar J and Panda S 2015 *ECS J. Sol. Stat. Sci. Technol.* 4 N18-N23
[29] Kumar S, Kumar N and Panda S 2016 *App. Biochem. Biotechnol.* DOI: 10.1007/s12010-016-1986-y
[30] Jang H-J, Ahn J, Kim M-G, Shin Y-B, Jeun M, Cho W-J and Lee K H 2015 *Biosens. Bioelectron.* 64 318-23

Journal of Physics: Conference Series **704** (2016) 012014 doi:10.1088/1742-6596/704/1/012014

Antibody conjugated graphene nanocomposites for pathogen detection

Chandan Sign[1, 2], Gajjala Sumana[1]

[1] Biomedical Instrumentation Section, Materials Physics and Engineering Division, National Physical Laboratory, (Council of Scientific & Industrial Research), Dr. K. S. Krishnan Road, New Delhi-110012, India

[2] Academy of Scientific and Innovative Research (AcSIR), CSIR-National Physical Laboratory Campus, Dr. K. S. Krishnan Marg, New Delhi-110012, India

E-mail: chandan.jadaun@gmail.com, sumanagajjala@gmail.com

Abstract. Graphene oxide (GO), due to its excellent electrochemical properties and large surface area, known to be highly suitable material for biosensing application. Here, we report *in situ* synthesis of silver nanopaticles (AgNPs) onto the GO sheets for the electrochemical detection of *Salmonella typhimurium* (*S.typhimurium*). The GO-AgNPs composites have been deposited onto the indium tin oxide (ITO) coated glass substrate by the electrophoretic deposition technique. Carbodiimide coupling (EDC-NHS) has been used for the immobilization of antibodies of *Salmonella typhimurium* (anti-*S.typhimurium*) for detection of *S.typhimurium*. The electron microscopy and UV-visible studies reveal successful synthesis GO-AgNPs composites while FT-IR studies suggest the proper immobilization of anti-*S.typhi*. The cyclic voltammetry (CV) has been utilized for detection using anti-*S.typhi*/GOAgNPs/ ITO based immunoelectrode as a function of *S.typhimurium* concentration. The fabricated immunosensor shows improved sensitivity of 33.04 μACFU^{-1}mLcm^{-2} in a wide detection range of 10^1 to 10^6 CFUmL^{-1}. This immunosensor may be utilized for the detection of other food borne pathogens like aflatoxin and *E.coli* also.

1. Introduction

The ability to detect various types of pathogen present in food supply chain like production, packaging and the storage is very important in order ensures the food safety and good health [1]. There are number of food borne diseases cause by bacterial and viral infection occurs across the globe. Food borne illness is responsible for 76 million illness, 325,000 hospitalizations and thousands death annually across the globe [2].Among various types of food borne pathogens, *S.typhimurium* is responsible for frequently occurring food borne illness. This pathogen infects various types of foods including chicken, ground beef, cucumber, and live poultry [3]. *S.typhimurium* is a gram-negative bacterium and origin of various types of illness including diarrhea, fever, and abdominal cramps [4]. It also generates localized infection in joints or even enters the human bloodstream. Several conventional methods have been developed for the detection of this toxic pathogen including specific agar media, to separate and count bacterial cells in particular sample. But these detection techniques take initially 3-4

[1] To whom any correspondence should be addressed.

days to generate the results and then further 6-7 days for the confirmation [5]. Since last two decades various electrochemical, electrical and optical detection methods have been reported for the detection of *S.typhimurium*. Among them electrochemical method found to be fast, reliable and affordable. Gehring *et al.* have reported the enzyme-linked immunomagnetic electrochemistry based technique for *S.typhimurium* detection [6]. Liebana *et al.* have developed the electrochemical detection technique based on a magneto-electrode [7].These techniques suffers several limitations including sensitivity and the detection limit. In this regard application of graphene oxide based traducers for the detection of *S.typhimurium* may overcome these challenges.

Since the introduction, graphene is considered as a highly suitable material for the biosensing application [8]. It is known to be a single layer of sp_2 bonded carbon atoms and exhibits a honeycomb structure having number of interesting properties such as very high electron mobility, room temperature quantum Hall effect; tunable optical properties, excellent electrical conductivity and high mechanical strength may provide potential applications in electrochemical detection of various bioanalytes [9-11]. Number of functional groups available on the basal and edge plane makes graphene oxide as a suitable material for biomolecular attachment [12]. Introduction of metal nanoparticles in the GO is known to improve the surface to volume ratio of the composites resulting into the improved loading of the biomolecules. Nanoparticles in host GO matrix improve the mechanical, catalytic and the conduction properties [13]. In addition to this, nanoparticles also act as the spacer between the GO sheets preventing them from the agglomeration. Silver nanoparticles (AgNPs) decorated GO is found to be highly suitable material for optoelectronics, catalysis and electrochemical applications. Wan *et al.* has designed a GO-AgNPs based efficient electrochemical biosensor to detect the anti-sulfate reducing bacteria [14]. A sensitive detection of tryptophan has also been reported by using GO-AgNPs transducer matrix [15]. In the current manuscript we have reported the *in situ* reduction and decoration of AgNPs in the GO sheet. This nanocomposite has been further utilized for the detection of *S.typhimurium* using CV technique.

2. Materials and Methods

2.1. Chemicals

Graphite powder flakes (45 μm, >99.99 wt %) were procured from Sigma Aldrich, USA. *S. typhimurium* antibody and antigens (heat killed *S. typhimurium* cells) were purchased from KPL Laboratory, USA. Bovine serum albumin (BSA), 1-ethyl-3-(3-dimethylaminopropyl) carbodiimide (EDC) and N-hydroxysuccinimide (NHS) were procured from Sigma, Aldrich, USA. Silver nitrate has been purchased from SD Fine Chem. Mumbai, India.

2.2. Characterization

The UV-visible spectroscopy (Perkin-Elmer, Lambda 950), Fourier transform infra-red spectroscopic (FT-IR, Perkin-Elmer, Model 2000) were used for characterization of the synthesized nanocomposite. The structural and morphological studies were carried out using scanning electron microscopy (SEM,LEO 440), transmission electron microscopy (TEM, Tecnaii-G2F30 STWIN). Electrochemical studies were performed on cyclic voltammetry conducted using Autolab, Potentiostat/Galvanostat (Eco Chemie, the Netherlands, Model AUT84275).

2.3. Methods

2.3.1. Synthesis of GO-AgNPs nanocomposite

Synthesis of GO has been achieved using the method reported in literature [16]. For the synthesis of GO-AgNPs, 1 ml of $AgNO_3$ (2 mM) has been added into the 10 mL of GO (1 mg/mL) kept in the smooth stirring with a temperature of 50ºC for 30 min. Then, 2 mL of ascorbic acid (5mM) has been added into the solution and the temperature has been raised to 70 ºC.

2.3.2. Electrophoretic deposition of GO-AgNPs on ITO electrode

Deposition of GO-AgNPs on the ITO electrode has been achieved using electrophoretic deposition (EPD) technique using two electrode systems. In details, using EPD setup 65V of DC voltage has been applied in the solution containing GO-AgNPs in the acetonitrile solution for 90 sec. In order to obtain a surface charge on the GO-AgNPs 10_{-5} to 10_{-4} mol of $Mg(NO_3)_2.6H_2O$ is added into the suspension as an electrolyte for EPD as Mg_{2+} ions adsorbed by the GO sheets provide an adequate positive surface charge to the GO-AgNPs sheets, resulting in enhanced deposition rate at the ITO electrode [17].

2.3.3. Biofunctionalization of the GO-AgNPs nanocomposite

Immobilization of the *anti-S.typhimurium* has been performed using carbodimide coupling chemistry. Prior to the covalent immobilization process, the COOH groups available in the GO-AgNPs electrode is activated, using EDC as the coupling agent and NHS as activator.10 µl of *anti-S.typhimurium* dissolved in the phosphate buffer (pH 7.4) has been uniformly dispersed onto the GO-AgNPs electrode and kept in the humid chamber for 6 hr. This allows the development of the strong covalent bonding between the COOH groups of the GO-AgNPs and the amine group of the anti-*S.typhimurium*. In addition to this, strong electrostatic interaction between the negatively charged AgNPs and the positively charged groups present in anti-*S.typhimurium* may also results into the higher loading of the anti-*S.typhimurium*. BSA, solution (1mg/mL) is used to block the non-specific active sites of the *anti-S.typhimurium*/GO-AgNPs bioelectrode.

3. Results and discussions

3.1. UV-Visible spectroscopy analysis

Figure 1A shows the detailed investigation of aqueous suspensions GO, AgNPs and GO-AgNPs using UV-Vis in the range of 200-800 nm with a slit width of 2 nm. The GO shows the peak at 231 nm mainly due to the π-π* transition of the various C-C bonds present in the GO. However, the composite shows a strong peak at 442 nm representing the significant number of AgNPs present in the nanocomposite.Another significant peak was found to be at 219 nm that may represent the blue shifting of the GO peak due to the introduction AgNPs. This phenomenon may occur due to the close conjugation of GO and the AgNPs resulting into the rapid electron transfer and increased transfer energy as reported in various literatures [18,19].

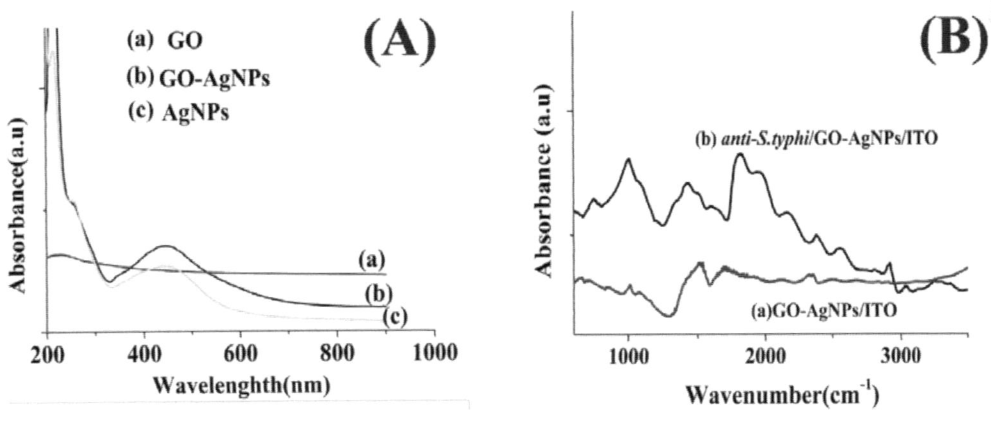

Figure 1. UV-visible studies of GO, AgNPs and GO-AgNPs (a), FTIR analysis of GO-AgNPs/ITO and *anti-S.typhimurium*/GO-AgNPs/ITO.

3.2. Furier Transform Infrared (FT-IR) spectroscopy analysis

FT-IR spectra of GO-AgNPs/ITO) and anti-*S.typhimurium*/GO-AgNPs/ITO films are shown in Figure

1B.The spectrum obtained for GO-AgNPs/ITO film shows a strong peak at 1598 cm^{-1} may be attributed to the stretching vibrations of C=C, the peak seen at 1297 $cm^{-1}$is due to C-O stretching. In the curve (b) strong peak at 1575 cm^{-1} represents the amide I bonding while the stretching, peak at 1256 cm^{-1}shows the C-O stretching and bending in the region of 3000-3500 cm^{-1} denotes primary and secondary amines representing the immobilization of *anti-S.typhimurium.*

3.3. Morphological Analysis

Figure 2 shows TEM images of the synthesized GO and the GO-AgNPs composites. Image (2A) shows the GO having various wrinkles and folding, while the image (2B) shows the AgNPs entirely distributed onto the GO sheets. The AgNPs, having different shape and size distributed on the GO sheet while, the AgNPs were in size range of 10-50 nm.

SEM investigation of GO-AgNPs and *anti-S.typhimurium*/GO-AgNPs/ITO immunoelectrode has been performed to get the surface characteristics of the synthesized materials (Fig. 3). Several agglomeration of the composite has been observed that may be due to the application of the high DC voltage for the fabrication of the film. Figure 3B shows the magnified small region on the GO-AgNPs as various layers of the GO composite with agglomerated AgNPs, are clearly visible. Figure 3C shows the EDX spectrum of the region shown in the image (3B). The GO-AgNPs composites have significant amount of AgNPs as shown in the image. After the immobilization of *anti-S.typhimurium* the GO-AgNPs/ITO, shows the uniform coverage of the biomolecule (Image 3D). This may be due to the strong interaction between the *anti-S.typhimurium* and the GO-AgNPs composites.

3.4. Electrochemical characterization studies

Electrochemical characterization of the GO-AgNPs and anti-*S.typhimurium*/GO-AgNPs immune electrode has been performed using CV (Figure 4A). The GO-AgNPs shows the anodic peak current of 322.2μA as shown in the curve (a). The higher amount of current may be due to AgNPs, as they offer the addition channels for the electron transport. But the *anti-S.typhimurium*/GO-AgNPs/ITO shows relatively lower value of the anodic peak current (212.04μA), as the immobilized biomolecules shows the insulation nature resulting into the blockage of the electron transport pathways.

The CV response of the *anti-S.typhimurium*/GO-AgNPs/ITO immunoelctrode as a function of scan rate (10-100 $mVsec^{-1}$) has been shown in the Figure 4B. It has been observed that the value of both anodic and cathodic peak current has been increases as a function of the square root of the scan rate indicating that redox reaction is controlled by semi-infinite linear diffusion process leading to the improved electron transfer kinetics. The slopes and the intercept are given by the following Eq. (1) and (2).

$$Ia[A] = 0.58 \times 10^{-6}[A] + 29.5 \times 10^{-6}\,[A^2s/mV]^{1/2} \times [scanrate\left(\frac{mV}{s}\right)]^{½};R^2=0.99 \quad (1)$$

$$Ic\,[A] = -34.55 \times 10^{-6}\,[A] - 19.60 \times 10^{-6}\,[A^2s/mV]^{1/2} \times [scanrate\left(\frac{mV}{s}\right)]^{½};R^2= 0.99 \quad (2)$$

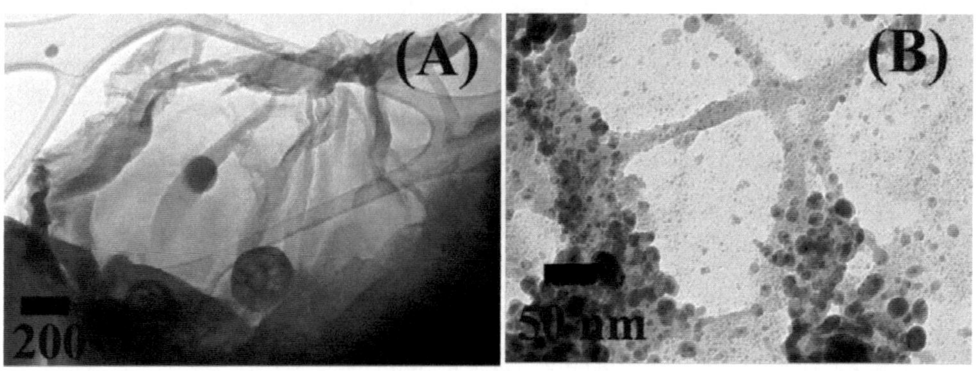

Figure 2. TEM image of synthesized GO (A), and GO-AgNPs nanocomposites (B)

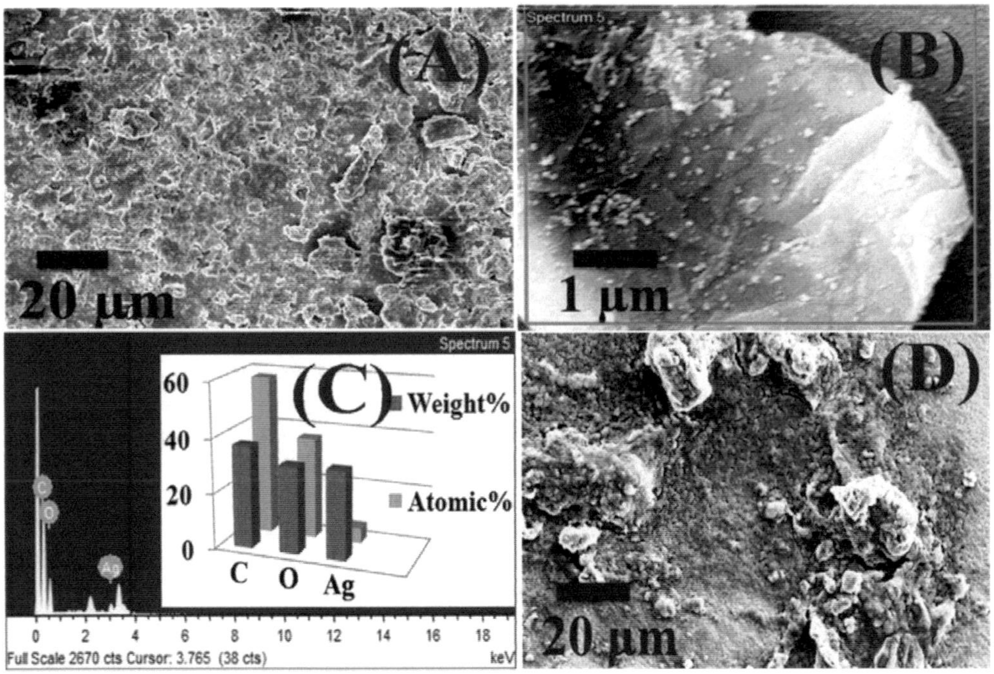

Figure 3. SEM image of GO-AgNPs/ITO (A), magnified region of GO-AgNPs/ITO (B) EDX spectrum of GO-AgNPs/ITO (C) and, *anti-S.typhimurium*/GO-AgNPs/ITO (D).

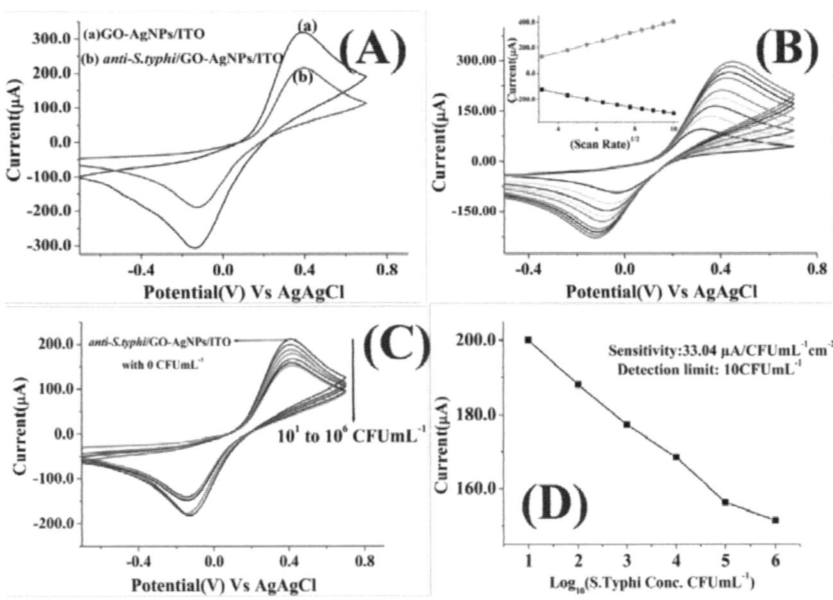

Figure 4. Electrochemical characterization of GO-AgNPs/ITO and anti-*S.typhimurium*/GOAgNPs/ ITO (A), Scan rate studies of anti-*S.typhimurium*/GO-AgNPs/ITO immunoelectrodes (B), Electrochemical response studies of anti-*S.typhimurium*/GO-AgNPs/ITO immunoelectrode as a function of *S.typhimurium* concentration (C), and Calibration plot obtained using anti-*S.typhimurium*/GO-AgNPs/ITO using different concentration of *S.typhimurium* (D).

3.5. Electrochemical response studies

Electrochemical response studies of anti-*S.typhimurium*/GO-AgNPs/ITO immunoelectrode has been performed as a function of *S.typhimurium* concentration (10^1 to 10^6 CFU/mL) using CV as shown in Figure 4C. It has been observed that the anodic peak current decreases as we increase the concentration of *S.typhimurium.* It may be due to the formation of the insulating antigen-antibody complex on the anti-*S.typhimurium*/GO-AgNPs/ITO immunoelectrode creating a barrier in the electron transport. As the concentration of the *S.typhimurium* increases the resistance gets increases in the same proportion. The calibration plot obtained between anodic peak current and *S.typhimurium* concentration has been shown in the Figure 4D.The fabricated immunosensor shows the high sensitivity of 33.04 $\mu A/CFU^{-1}mLcm^{-2}$and the lower detection limit of 10 $CFUmL^{-1}$.The higher loading of anti-*S.typhimurium* through the strong amide linkage and the electrostatic interaction with the composite may results into the strong signals. Superior loading of the anti-*S.typhimurium* using the carbodimide coupling may have resulted in the capturing of the higher number of incoming *S.typhimurium* cells. In addition to this AgNPs present on the surface of GO may have offers additional sites for the loading of anti-*S.typhimurium*by as electrostatic interaction between the negatively charged AgNPs and the positively charged amine ($-NH_2$) group available in the anti-*S.typhimurium.*

The designed immunosensor did not show any significance interference in the presence of other potential pathogens like *E.coli*. As shown in the Figure 5A only a negligible change (2.8%) in the current value has been observed, when *E.coli* has been added into the *S.typhimurium* (10^4 $CFUmL^{-1}$). In addition to this the anti-*S.typhimurium*/GO-Ag/ITO immunosensor found to be reproducible as shown in Figure 5B. Six different anti-*S.typhimurium*/GO-AgNPs/ITO immunoelectrode using 10^4 $CFUmL^{-1}$ of *S.typhimurium*concentration shows almost same current.These immunoelectrodes has been prepared under the same condition and did not shows any significant change in the current value. The mean value has been calculated as 168.2 µA.

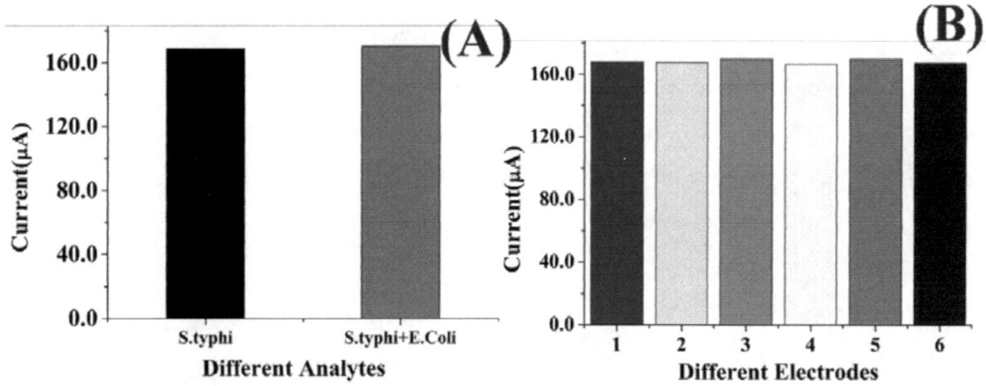

Figure 5. Interference study of anti-*S.typhimurium*/GO-AgNPs/ITO immunosensor (A), repeatability studies of anti-*S.typhimurium*/GO-AgNPs/ITO (B).

4. Conclusion

The GO-AgNPs based electrochemical immunosensor for *S.typhimurium* detection has been investigated in the present manuscript. A natural green and eco friendly route has been developed for the synthesis of GO-AgNPs using ascorbic acid. A well dispersed GO-AgNPs in acetonitrile has been utilized for the electrophoretic deposition and further EDC-NHS chemistry has been used to attach anti-*S.typhimurium.* Electron microscopy, and UV studies reveal successful synthesis of GO-AgNPs whereas FT-IR studies suggest its proper functionalized with anti-*S.typhimurium.* The fabricated immunosensor found to be efficient for the detection of *S.typhimurium* and may be utilized for other water and food borne pathogens.

Acknowlegement
We are highly thankful to Director, CSIR-National Physical Laboratory, New Delhi, India for providing the facilities. Chandan Singh is thankful CSIR, India for providing financial support as Senior Research Fellowship. Authors are thankful to Mr. Dinesh Singh and Jai Talwade for TEM and SEM investigations respectively. Authors are also thankful to Mr. Manoj Kumar Pandey for the technical support. Financial support received from the ESC-0103 is highly acknowledged.

Reference

[1] Chai Y, Li S, Horikawa, S, Park M K, Vodyanoy V, and Chin B A2012, *J. Food Prot.* , **75**, 631
[2] Guntupalli R, Lakshmanan R S, Johnson M L, Hu J, Huang T S, Barbaree J M, and Chin B A, 2007, *Sens Instrum Food QualSaf*, **1**, 3
[3] Onoda Ghosh D S, Barizuddin S, Yuksek N S, Dweik M, and Almasri M F, 2015. *J. Sens.* http://dx.doi.org/10.1155/2015/293461
[4] Scharff R L, 2012. *J. Food Prot.*, **75**, 123
[5] United States Department of Agriculture, Fact Sheets: Foodborne Illness and Disease, 2011, http://www.fsis.usda.gov/wps/portal/fsis/topics/food-safety-education/get-answers/foodsafety-fact-sheets/foodborne-illness-and-disease/salmonella-questions-and-answers/ct_index.
[6] Gehring A G, Crawford C G, Mazenko R S, Van Houten L J, and Brewster J D. 1996,J. Immunol. Methods, **195**, 15
[7] Liébana S., Lermo A, Campoy S, Cortés M P, Alegret S, and Pividori M I,2009. *Biosens. Bioelectron.*, **25**, 510.
[8] Kim J, Kim F, and Huang J., 2010, *Mater. Today*, **13**, 28
[9] Huang X, Yin Z, Wu S, QiX, He Q, Zhang Q,and Zhang H. 2011, *Small*, **7**, 1876
[10] Wassei J K, andKaner R B, 2010, *Mater. Today*, **13**,52
[11] Huang Y, Dong X, Shi Y, Li C M, Li L J, and Chen P, 2010 *Nanoscale*, **2**, 1485
[12] Srivastava S, Kumar V, Ali, M A, Solanki P R, Srivastava A, Sumana G, and Malhotra B D, 2013, *Nanoscale*, **5**, 3043
[13] Dhand V, Kyong Y R, Hyun J K, and Dong H J, 2013, *J. Nanomater.***2013**, 158
[14] Wan Yi, Yi W, Jiajia W, and Dun Z, 2010, *Anal. Chem.* **83**, 648
[15] Li J, Daizhi K, Yonglan F, Fuxing Z, Zhifeng X, Mengqin L, and Deping W, 2013, *Biosens. Bioelectron.***42**, 198
[16] Marcano D C, Kosynkin D V, Berlin J M, Sinitskii A, Sun Z, Slesarev A, and Tour J M, 2010, *ACS Nano*, **4**, 4806
[17] Singh C, Srivastava S, Ali M A, Gupta T K, Sumana G, Srivastava A, and Malhotra B D, 2013, *Sens. Actuators, B*, **185**, 258
[18] Zhong L, and Kyusik Y, *Int. J. Nanomed.***10**, 79
[19] Sahu S R, Mayanglambam M D, Puspal M, Pratik S, and Krishanu B. 2013, *J. Nanomater.*, **6**, 65

DNA methylation detection based on difference of base content

Shinobu Sato,[1,2] Keiichi Ohtsuka,[1,2] Satoshi Honda,[1] Yusuke Sato[1] and Shigeori Takenaka[1,2,*]

[1]Department of Applied Chemistry, Kyushu Institute of Technology, Kitakyushu, Japan

[2]Research Center for Bio-microsensing Technology, Kyushu Institute of Technology, Kitakyushu, Japan

shige@che.kyutech.ac.jp

Abstract. Methylation frequently occurs in cytosines of CpG sites to regulate gene expression. The identification of aberrant methylation of certain genes is important for cancer marker analysis. The aim of this study was to determine the methylation frequency in DNA samples of unknown length and/or concentration. Unmethylated cytosine is known to be converted to thymine following bisulfite treatment and subsequent PCR. For this reason, the AT content in DNA increases with an increasing number of methylation sites. In this study, the fluorescein-carrying bis-acridinyl peptide (FKA) molecule was used for the detection of methylation frequency. FKA contains fluorescein and two acridine moieties, which together allow for the determination of the AT content of double-stranded DNA fragments. Methylated and unmethylated human genomes were subjected to bisulfide treatment and subsequent PCR using primers specific for the CFTR, CDH4, DBC1, and NPY genes. The AT content in the resulting PCR products was estimated by FKA, and AT content estimations were found to be in good agreement with those determined by DNA sequencing. This newly developed method may be useful for determining methylation frequencies of many PCR products by measuring the fluorescence in samples excited at two different wavelengths.

1. Introduction

Cancer-suppressing genes are inactivated by the aberrant methylation of cytosine in the CpG islands of their promoter regions [1]. Such inactivation leads to cancer via loss of physiological function of these genes or due to a loss of transcription factor activity resulting from secondary methylation of related genes [2]. The detection of methylation frequency in cancer-suppressing genes has thus become an important aspect of cancer diagnosis. Cytosine methylation can be detected by (1) bisulfite sequencing [3], (2) methylation-specific PCR [4], or (3) combined bisulfite restriction analysis (COBRA) methods [5]. These methods, however, have limitations such as time-consuming procedures with tedious steps and the formation of unclear internally formed PCR products or DNA sequencing capacity. To overcome these limitations, several improved methods have been developed [6-13]. To overcome the limitations of method 1, direct detection of methylated cytosine was assessed without bisulfite treatment. Okamoto and co-workers succeeded in detecting methylated cytosine based on the specific reaction with osmium complex [6], while Niwa and co-workers were able to discriminate between methylated and unmethylated cytosine electrochemically based on differences in redox potentials [7]. Another approach was to use surface plasmon resonance (SPR) with a methylated cytosine-specific antibody [8]. Since unmethylated cytosine is converted to thymine after bisulfite treatment and subsequent PCR, the AT content is increased with increasing methylation frequency. Accordingly, methylation frequency can be determined based on T_m values of DNA compared with unmethylated DNA [9]. However, this method has a limitation that the amplification region for PCR analysis is subject to T_m value limitations, where higher T_m values are not measured accurately. The limitations associated with methods 2 and 3 were addressed by combining DNA hybridization after bisulfite treatment with subsequent PCR. We achieved electrochemical methylation detection in PCR products after bisulfite treatment coupled with DNA-immobilized electrode and ferrocenylnaphthalene diimide (FND) as a double-stranded DNA-specific ligand [10-13]. High sensitivity detection of the hybrid species was achieved using a surface-enhanced Raman scattering method [14]. Primer and target regions are discriminated precisely in these methods. Here, a bis-acridinyl peptide carrying fluorescein (FKA, Fig. 1A) [15] was applied to estimate methylation frequency in PCR products, where FKA exhibits almost no fluorescence in aqueous solutions because of intramolecular quenching by fluorescein being sandwiched between two acridine moieties. When FKA is bound to double-stranded DNA, the fluorescein of FKA is eliminated by the intercalation of the acridine portion of the peptide between the base pairs of the DNA duplex and thus the fluorescence is recovered. Interestingly, the fluorescence of fluorescein (λ_{em} = 517 nm) was observed via FRET when excited at 422 nm based on the acridine chromophore and this fluorescence correlated with AT content, since the acridine fluorescence was quenched due to binding to GC base pairs (Fig. 1B). The fluorescence (517 nm) obtained with an excitation wavelength of 492 nm, however, correlated with the amount of DNA present (Fig. 1B). The AT content of DNA was therefore estimated by the fluorescence ratio of $\Delta FI_{422}:\Delta FI_{492}$ ($\Delta FI_{422}/\Delta FI_{492}$) [15]. Since unmethylated cytosine is converted to thymine after bisulfite treatment and PCR, AT content differs between methylated and unmethylated PCR products. The AT content in DNA can be detected by FKA without knowing the DNA concentration and thus we hypothesized that FKA can be used to estimate the AT content of PCR products of unknown length and sequence. To confirm this hypothesis, the methylation frequencies in PCR products representing portions of the *CDH4* [16], *CFTR* [17], *DBC1* [17], and *NPY* [17] genes—all with different lengths and sequences—were assessed using FKA.

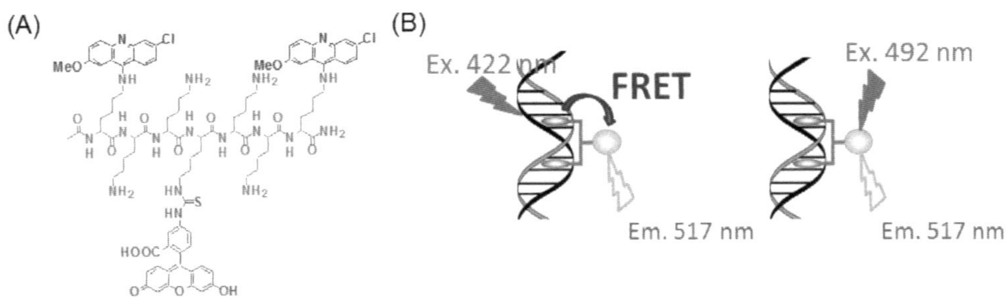

Fig. 1. (A) Chemical structure of the fluorometric peptide-type intercalator FKA and (B) the principle of AT content estimation with FKA.

2. Experimental

2.1. Optimization of experimental conditions

Fluorescence measurements were carried out using 96-well titer plates (Nunc) containing 100 μL/well of either blank solution (0.5 μM FKA, 5 mM Tris-HCl [pH 8.3], 0.5 mM EDTA) or test solution (20

μM (-bp) dsDNA, 0.5 μM FKA, 5 mM Tris-HCl (pH 8.3), and 0.5 mM EDTA) as well as 100 mM, 200 mM, or 400 mM NaCl. A Molecular Devices SPECTRA MAX GEMINI XS instrument was used with excitation wavelengths of either 422 nm or 492 nm and an emission wavelength of 517 nm. Double-stranded DNA species with different AT contents were used: poly[(dA-dT)]$_2$, *Clostridium perfringens* DNA, calf thymus DNA, *Escherichia coli* DNA, *Micrococcus lysodeikticus* DNA, and poly[(dG-dC)]$_2$, which have known AT contents of 100%, 73%, 57%, 28%, and 0%, respectively.
The fluorescence readings obtained were evaluated using equations 1, 2, and 3 below, where $FI_{ex.422, DNA}$ and $FI_{ex.422, blank}$ represent fluorescence at 517 nm when excited at 422 nm in the presence (DNA) or absence (blank) of DNA duplex, respectively; and $FI_{ex.492, DNA}$ and $FI_{ex.492, blank}$ represent fluorescence at 517 nm when excited at 492 nm in the presence (DNA) or absence (blank) of DNA duplex.

$$\Delta FI_{422} = FI_{ex.422, DNA} - FI_{ex.422, blank} \quad (1)$$

$$\Delta FI_{492} = FI_{ex.492, DNA} - FI_{ex.492, blank} \quad (2)$$

From (1) and (2), $\text{F-Ratio} = \Delta FI_{422}/\Delta FI_{492}$ (3)

2.2 Material collection

Oligonucleotide template genes or PCR primers (Table 1) were purchased from Genenet (Fukuoka, Japan). The template DNA sequences of the *CFTR* gene were as follows [16]: methylated *CFTR* gene (CFTR-M), 5′-<u>GTT TTG GGT TTG GCG GAT TTT GA</u>C GCG AAG GAG GGT TTA GGA AGT TTT TCG GGG AGT CGG TTT TTT CGT CGG TGG TTT TTT TTG TTT TTT AGC GTT GTT AAT TGG ATT TAA AGA GAG <u>GTC GGA TTG TCG TTT ATT TGC GGG</u>-3′, unmethylated *CFTR* gene (CFRT-U): 5′-<u>GGT TTT GGG TTT GGT GGA TTT TGA TGT</u> GAA GGA GGG TTT AGG AAG TTT TTT GGG GAG TTG GTT TTT TTG TTG GTG GTT TTT TTT GTT TTT TAG TGT TGT TAA GAT TTA AAG AGA G<u>GT TGT GAT TGT TGT TTA TTT GTG GGA TG</u>-3′. CpGenome™ Universal Methylated DNA (Lot PSO 1566253, Merck Millipore) and CpGenome™ Universal Unmethylated DNA (Lot Number PSO 1579958, Merck Millipore) were used as methylated and unmethylated human genome DNA, respectively. The PrimeStar (TAKARA Bio) DNA polymerase was used to amplify oligonucleotide PCR templates, while ZymoTaq PreMix (Zymo research) was used to amplify genomic DNA after bisulfite treatment.

2.3. Bisulfite treatment

CpGenome™ Universal Methylated DNA (Lot PSO 1566253) and CpGenome™ Universal Unmethylated DNA (Lot Number PSO 1579958) were used as methylated and unmethylated samples, respectively. The above genomic DNA species (1.0 μg) were treated with the CpGenome™ FAST DNA Modification Kit (Merck Millipore) according to the manufacturer's instructions. The concentrations of the bisulfite-treated human genome templates were estimated by Nanodrop (Thermo Scientific) to be 22.9 ng/μl (Methylated, Lot PSO 1566253) and 17.4 ng/μl (Unmethylated, Lot Number PSO 1579958).

2.4. Polymerase Chain Reaction (PCR)

Table 1 shows the sequences of the primers used in PCR. Total reaction volumes of 100 μL were prepared: 4.0 μL CFTR-M (1 nM) or CFTR-U (1 nM) (or 1.0 μL of 22.9 or 17.4 ng/μL template DNA), 4 μL each of F- and R-primers (10 μM), and 50 μL PrimeStar (or ZymoTaq). PCR was carried out under the following conditions: 95°C for 5 min, followed by 30 cycles of 94°C for 15 sec, annealing temperature [different for each template] for 5 sec, and 72°C for 10 sec. The resulting PCR products were purified using a QIAquick PCR Purification Kit (QIAGEN) and were eluted in 30 μl elution buffer. The purified PCR products were assessed by 12.5% native polyacrylamide gel electrophoresis (PAGE).

2.5. PCR product sequencing

Sequencing of PCR products was carried out by Hokkaido System Science (Sapporo, Japan).

Journal of Physics: Conference Series **704** (2016) 012015 doi:10.1088/1742-6596/704/1/012015

2.6. Fluorescence analysis of PCR products

Fluorescence measurements were carried out in microtiter plates containing 100 μL/well of the following solution: 0.6 μM-bp PCR product, 0.5 μM FKA, 5 mM Tris-HCl (pH 8.3), and 0.5 mM EDTA as well as 100 mM, 200 mM, or 400 mM NaCl using a Molecular Devices SPECTRA MAX GEMINI XS instrument with excitation wavelengths of 422 nm or 492 nm and an emission wavelength of 517 nm. The resulting fluorescence values were evaluated using equation (3).

Table 1. Primer sequences used in this experiment

Name	Sequence 5'→3'	Product size (bp)	Tm (°C)	PCR Cycles
CFTR_MF	GTTTTGGGTTTGGCGGATTTTGACGC	142	46 56*	40 30*
CFTR_MR	CCCGCAAATAAACGACAATCGCGAC			
CFTR_UF	GGTTTTGGGTTTGGTGGATTTTGATGT	146	46 56*	40 30*
CFTR_UF	CATCCCACAAATAAACAACAATCACAAC			
CDH4_MF1	TTGTAGTTTCGAGCGCGC	117	45	40
CDH4_MR1	CCCGACTCCGAAAACCAAA			
CDH4_MF2	GTTTTCGGTGTCGGGTATC	105	45	40
CDH4_MR2	CGACAACTTACCCGAAACG			
DBC1_MF	ACGCGATCCCTTTAAATACTCGTACG	131	60	40
DBC1_MR	GAGGAGAGACGGGAGGTCGTTTCG			
NPY_MF	TTAAAACCCTCTAACCGAAAACTTCCG	131	45	40
NPY_MR	ACGATTAGCGCGGTATTTTCGTCGG			

*Custom Synthesized Template oligonucleotide

Results and Discussion

3.1 Optimization of detection conditions

Fluorescence at 515 nm was measured for 20–50 μM-bp $[poly(dA\text{-}dT)]_2$, calf Thymus DNA, or $[poly(dG\text{-}dC)]_2$ using 0.50 μM FKA in 5 mM Tris-HCl (pH 8.3), 0.5 mM EDTA, and 200 mM NaCl excited at 422 nm or 492 nm and evaluated in terms of F-Ratio calculated by equation (3). F-Ratios were shown to differ depending on AT content, but were found to be stable at different DNA concentrations (Fig. 2).

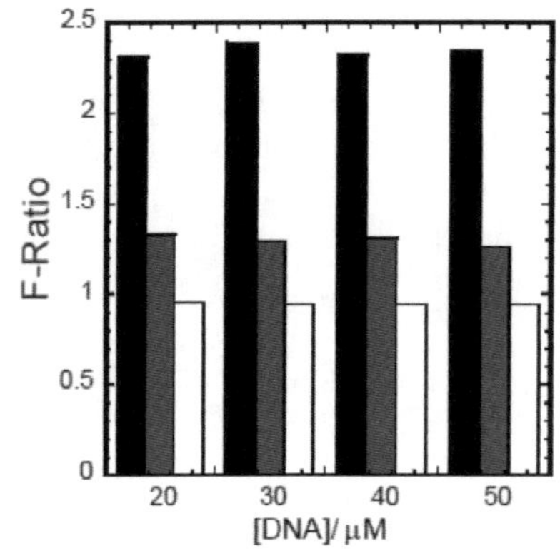

Fig. 2. Estimation of AT content for DNA samples of unknown concentration. Black bar: [poly(A-T)]$_2$, gray bar: calf thymus DNA, white bar: [poly(dG-dC)]$_2$. Experiments were carried out for 20–50 μM-bp DNA in 5 mM Tris-HCl (pH 8.3), 0.5 mM EDTA, and 200 mM NaCl containing 0.5 μM FKA.

Here, six different PCR products were prepared using four different genes (Table 1). These PCR products were fixed at a concentration of 0.6 μM per double-stranded DNA, while the PCR products of 105–146 bp in length corresponded to concentrations of 63–88 μM-bp and were thus subjected to AT content quantification using F-Ratio. In the next step, the effect of salt concentration on AT content estimation by FKA was assessed with NaCl concentrations of 100, 200, or 400 mM using poly[(dA-dT)]$_2$, *C. perfringens* DNA, calf thymus DNA, *E. coli* DNA, *M. lysodeikticus* DNA, and poly[(dG-dC)]$_2$ with AT contents of 100%, 73%, 57%, 28%, and 0%, respectively. Figure 3 shows that the F-Ratios varied with AT content and that the greatest change in F-Ratio was observed with 400 mM NaCl for DNA with >50% AT content. Since these PCR products had AT contents of 37.8–65.1%, subsequent experiments were carried out with 400 mM NaCl.

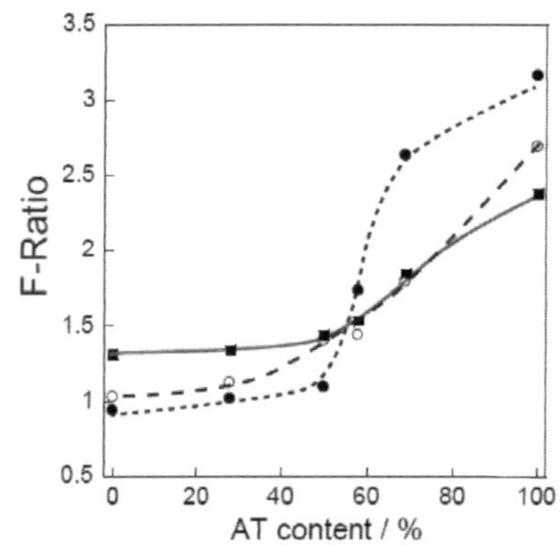

Figure 3. F-Ratio depends on AT content and NaCl concentration. Experiments were carried out with 20 μM DNA, 5 mM Tris-HCl (pH 8.3), 0.5 mM EDTA, 0.5 μM FKA, and NaCl at 100 (■), 200 (○), or 400 (●) mM.

3.2 AT content estimation using the model system

As an initial trial use of the newly developed model system, AT content was estimated by FKA for assessing methylation frequency in the *CFTR* gene, the methylation of which is associated with bowel cancer [17]. The CFTR-M (AT content: 65.1%) and CFTR-U (AT content: 56.8%) DNA species were used as methylated and unmethylated *CFTR* genes following bisulfite treatment, after which PCR products were prepared using these DNA species as templates. A single band of ~140 bp in size was

Journal of Physics: Conference Series **704** (2016) 012015 doi:10.1088/1742-6596/704/1/012015

obtained in native PAGE analysis of the PCR products (data not shown). F-Ratios of these PCR products were determined using FKA and were found to be 1.65 and 1.05 for the methylated and unmethylated samples, respectively, which is in agreement with the data shown in Fig. 3. The CFTR-M and CFTR-U DNA species were combined at ratios of 0:1, 0.25:0.75, 0.5:0.5, 0.75:0.25, and 1:0 ratio to yield AT contents of 65.1, 63.0, 61.0, 58.9, and 56.8%, respectively. The resulting F-Ratios are shown in Fig. 4 and were found to correlate well with the theoretical values (Fig. 3).

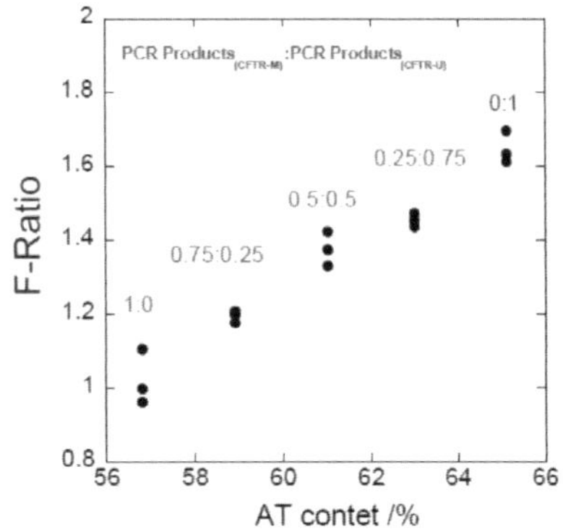

Figure 4. Detection of methylation frequency in the *CFTR* gene by FKA. Experiments were carried out with 0.6 μM PCR product in 5 mM Tris-HCl (pH 8.3), 0.5 mM EDTA, and 400 mM NaCl containing 0.5 μM FKA.

3.3 Methylation detection in genomic samples

The *CFTR, CDH4, DBC1, and NPY* genes were subjected to PCR after bisulfite treatment. CpGenome™ Universal Methylated DNA (Lot PSO 1566253) and CpGenome™ Universal Unmethylated DNA (Lot Number PSO 1579958) were used as template DNA and the oligonucleotides listed in Table 1 were used as primers. The *CFTR, CDH4, DBC1,* and *NPY* genes correlate with bladder cancer, large intestinal cancer, lung cancer (non-small cell), and neuropeptide Y (a factor indirectly causing cancer), respectively. The resulting PCR products each yielded single bands of the expected sizes in native PAGE (data not shown). Although the methylation frequencies and AT contents of these products are not known, they were determined by DNA sequencing (Fig. A1). The AT contents of the PCR products are listed in Table 2, where the DNA sequencing results are shown together with the theoretical values (assuming completely methylated genomic DNA). The PCR products were then tested using the 0.5 μM FKA method and the resulting F-Ratios were plotted against AT content (Fig. 5). The resulting F-Ratios were in good agreement with the expected AT content, yielding good correlation between the two parameters.

Table 2. AT content determined from DNA sequencing compared with theoretical values.

AT content (%)	CFTR-M	CFTR-U	CDH4 Region 1	CDH4 Region 2	DBC1	NPY
Theoretical	56.8	65.1	37.8	42.9	43.5	59.5
Based on sequencing	56.8	63.0	37.8	51.4	43.5	59.5

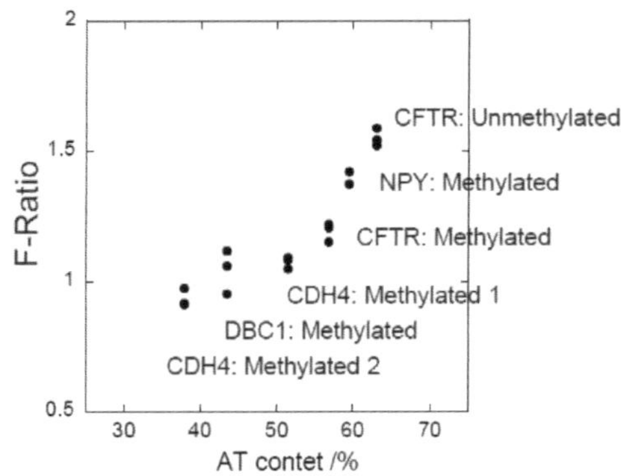

Fig. 5. Relationship between F-Ratio and AT content of methylated and unmethylated PCR products after bisulfite treatment. Experiments were carried out with 0.6 μM PCR product in 5 mM Tris-HCl (pH 8.3), 0.5 mM EDTA, and 400 mM NaCl containing 0.5 μM FKA.

4. Conclusion

Methylation frequency based on AT content was estimated by FKA for the *CFTR, CDH4, DBC1*, and *NPY* genes amplified from human genome samples subjected to bisulfite treatment. This method does not allow for the location of methylated site of PCR products to be determined; however, methylation frequency was estimated in double-stranded DNA samples of different lengths and/or unknown concentrations using this simple procedure. This newly developed method may be useful in determining methylation frequencies of many PCR products by simply measuring the fluorescence in samples excited at two different wavelengths. Furthermore, it is possible to estimate the methylation frequency in a DNA sample via real-time methods using an RT-PCR instrument and carrying out the reactions in the presence of FKA after bisulfite treatment of sample DNA. For this method to be effective in real-time, one would need to ensure that the polymerase is not inhibited during the PCR stage of the process.

References

[1] Ohtani N, Fujita T, Aoike A, Osifchin N, Robbins P and Sakai T 1993 *Oncogene*, **8** 1063.
[2] Abe M, Ohira M, Kaneda A, Yagi Y, Yamamoto S, Kitano Y, Takato T, Nakagawara A and Ushijima T 2005 *Cancer Res.* **65** 828.
[3] Frommer M, Mcdonald L E, Millar D S, Collis C M, Watt F, Grigg G W, Molloy P L and Paul C L 1992 *Proc. Natl. Acad. Sci. USA* **89** 1827.
[4] Herman J G, Graff J R, Myöhänen S, Nelkin B D and Baylin S B 1996 *Proc. Natl. Acad. Sci. USA* **93** 9821.
[5] Miotto E, Sabbioni S, Veronese A, Calin G A, Gullini S, Liboni A, Gramantieri L, Bolondi L, Ferrazzi E, Gafa R, Lanza G and Negrini M 2004 *Cancer Res.* **64** 8156.
[6] Tanaka K, Tainaka K, Umemoto T, Nomura A and Okamoto A 2007 J. Am. Chem. Soc. **129**14511.
[7] Kurita R and Niwa O 2012 *Anal. Chem.* **84** 7533.
[8] Kato D, Sekioka N, Ueda A, Kurita R, Hirono S, Suzuki K, Niwa O 2008 *J. Am. Chem. Soc.* **130** 3716.
[9] Guldberg P, Worm J, Grønbaek K 2002 *Methods* **27** 121.
[10] Sato S, Kondo H, Hokazono K, Irie T, Ueki T, Waki M, Nojima T and Takenaka S 2006, *Anal. Chim. Acta*, **578** 82.
[11] Sato S, Tsueda M and Takenaka S 2010 *J. Organomet. Chem.* **695** 1858.
[12] Sato S, Tsueda M, Kanezaki Y and Takenaka S 2012 *Anal. Chim. Acta.* **715**, 42.

[13] Sato S, Saeki T, Tanaka T, Kanezaki Y, Hayakawa M, Haraguchi K, Kodama M, Nishihara T, Tominaga K and Takenaka S 2014 *Appl. Biochem. Biotech.* **174** 869.
[14] Wang Y, Wee E J H and Trau M 2015 *Chem. Commun.* **51** 10953.
[15] Ueyama U, Mizuki K, Nojima T and Takenaka S 2004 *Analyst* **129** 886.
[16] Bibikova M, Lin Z, Zhou L, Chudin E, Garcia E W, Wu B, Doucet D, Thomas N J, Wang Y, Vollmer E, Goldmann T, Seifart C, Jiang W, Barker DL, Chee M S, Floros J and Fan J B 2006 *Genome Res.* **16** 383.
[17] Miotto E, Sabbioni S, Veronese A, Calin GA, Gullini S, Liboni A, Gramantieri L, Bolondi L, Ferrazzi E, Gafà R, Lanza G and Negrini M 2004 *Cancer Res.* **64** 8156.

Acknowledgments
This work was supported in part by Grants-in-Aid for Scientific Research from the Ministry of Education, Culture, Sports, Science and Technology (MEXT).

Influence of emissive layer thickness on electrical characteristics of polyfluorene copolymer based polymer light emitting diodes

D Das[1], P Gopikrishna[1], A Singh[1], A Dey[1] and P K Iyer[1, 2, *]

[1]Centre for Nanotechnology, Indian Institute of Technology Guwahati, Guwahati 781039, Assam, India
[2]Department of Chemistry, Indian Institute of Technology Guwahati, Guwahati 781039, Assam, India

[*]pki@iitg.ernet.in

Abstract. Polymer light emitting diodes (PLEDs) with a device configuration of ITO/ PEDOT:PSS/ PFONPN01 [Poly[2,7-(9,9'-dioctylfluorene)-co-N-phenyl-1,8-naphthalimide (99:01)]/LiF/Al have been fabricated by varying the emissive layer (EML) thickness (40/65/80/130 nm) and the influence of EML thickness on the electrical characteristics of PLED has been studied. PLED can be modelled as a simple combination of resistors and capacitors. The impedance spectroscopy analysis showed that the devices with different EML thickness had different values of parallel resistance (R_P) and the parallel capacitance (C_P). The impedance of the devices is found to increase with increasing EML thickness resulting in an increase in the driving voltage. The device with an emissive layer thickness of 80nm, spin coated from a solution of concentration 15 mg/mL is found to give the best device performance with a maximum brightness value of 5226 cd/m^2.

1. Introduction

Polymer light-emitting diodes (PLEDs) have been a major area of attraction for the researcher over past few decades due to its potential application in the field of flat-panel displays and solid state lighting sources due to their low cost, ease of fabrication, brightness, speed, wide viewing angle, low power consumption and contrast [1-5]. The performances of such device are critically dependent on the properties of both emissive materials and their processing parameters. Among PLEDs, stable and pure blue light emitting devices is of utmost importance as it plays a vital role in realizing full color displays and high quality white light [6,7]. Polyfluorenes (PFs) are well-known emissive material for blue polymer light emitting diodes (PLEDs) due to their high luminescence efficiency, excellent thermal stability, charge transport property, good processability and color tunability [8,9]. To achieve high efficiency for a given polymer, it is necessary to balance and confine electrons and holes inside the emissive layer [10]. Unfortunately, in case of PFs the hole mobility is much larger than the electron mobility. However, adding electron withdrawing moieties like 1,8 napthalamide (NPN) into the polyfluorene main chain can significantly enhance its electron transporting properties[11,12].

In this paper, we report the fabrication of blue PLEDs using PFONPN01 [Poly[2,7-(9,9'-dioctylfluorene)-co-N-phenyl-1,8-naphthalimide (99:01)] as emissive layer (EML) and the effect of EML thickness on the electrical parametres of PLEDs. The impedance of the devices is found to increase with increasing EML thickness resulting in an increase in the driving voltage. The device with an EML thickness of 80nm, spin coated from a solution of concentration 15 mg/mL is found to give the best device performance with a maximum brightness value of 5226 cd/m^2.

2. Experimental Section

2.1. *Chemicals and substrate*

Indium tin oxide (ITO) coated glass substrate (sheet resistance 15 Ω/sq.), Poly(3,4-ethylenedioxythiophene)-poly(styrenesulfonate) (PEDOT: PSS), Lithium fluoride (LiF) and Aluminum wire (99.999% purity) were purchased from Sigma Aldrich and used as received. PFONPN01 used as emissive layer was synthesized using previously reported method [11] and its chemical structure is shown in figure 1(a).

2.2. *Characterization details*

The thin film of PEDOT:PSS and PFONPN01 were deposited on ITO coated glass substrate by spin coating technique using Laurell and Spin 150 spin coater outside and inside the Jacomax glove box respectively. The Veeco Dektak 150 Surface Profilometer was used to measure thicknesses of the thin films. All current density vs voltage (J-V) characteristics of the fabricated devices were measured by Keithley-2400 digital source meter. The luminance values of the PLEDs were measured using LCS 100 integrated sphere. CH 680 Instrument was used to measure the impedance spectroscopies of all the devices. All the device characterizations were carried out inside glove box under inert gas environment.

2.3. *Device Fabrication Method*

PLEDs were fabricated using PFONPN01 copolymer as emissive layers. The PLED device configuration is composed of a pre-cleaned and pre-patterned ITO as the transparent anode. The ITO surface was ultrasonically agitated and cleaned in 2% detergent, acetone and isopropanol, each for ten minutes. After cleaning, an 40 nm thin layer of poly (3,4-ethylendioxythiophene): poly(styrenesulfonate) (PEDOT:PSS) as a hole injecting layer was spin-coated at 3000 rpm for 60 seconds, then baked for 15 minutes in an argon environment at 130°C. The copolymers were dissolved in chloroform at four different concentration (5/10/15/20 mg/mL) and spin-coated above the PEDOT:PSS layer under an ambient atmosphere and thermally treated at 80°C for 15 minutes for the removal of residual solvent. The film thickness of the emissive layer was~ 80 nm. Finally, LiF and aluminum was thermally evaporated at a rate of 0.1 Å s^{-1} and 10 Å s^{-1} respectively at a base pressure of 10^{-6} mbar to form the cathode electrode. The emissive area of the diodes was 20 mm^2. The schematic device structure is shown in figure 1(b).

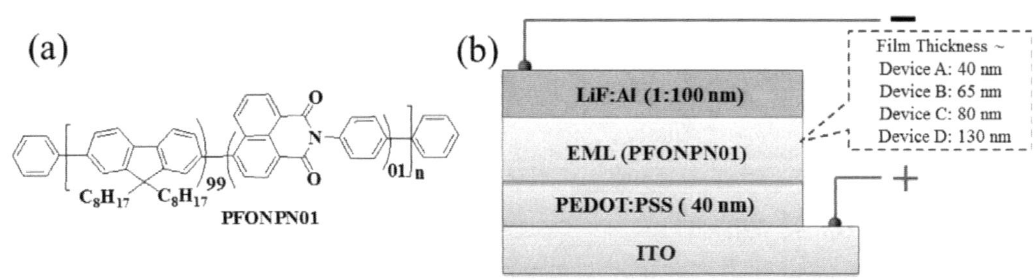

Figure 1. (a) Chemical structure of PFONPN01 (b) Schematic device structure

3. Result and Discussion

The impedance of the fabricated devices with different EML thicknesses were measured using CH680 instrument in the range of 10 Hz to 1 MHz at zero dc bias voltage. The amplitude of the ac test signal was 50 mV. The results of the measurements were analyzed based on the complex impedance equation:

$$Z = \frac{1}{Y} = R + jX = R + \frac{1}{jwC} = Z' + jZ''$$

where Z, Y , R, X and C are the impedance, admittance, resistance, reactance, and capacitance of the device, respectively.

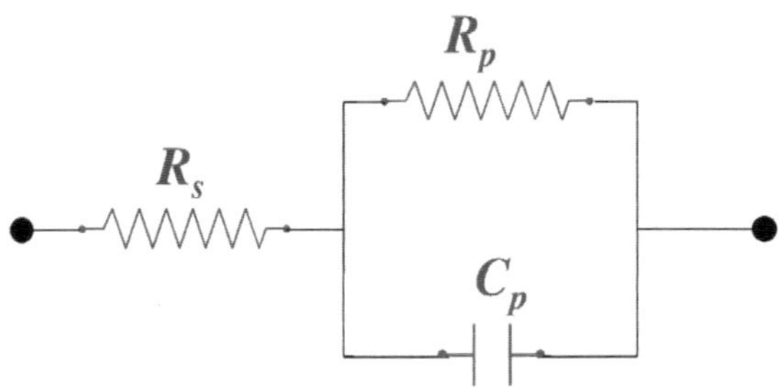

Figure 2. Equivalent circuit model for PLED

Table 1. Comparison of the resistance and the capacitance values for the devices with different emissive layer thickness

Device Name	EML thickness (nm)	R_S (ohm)	R_p (ohm)	C_p (F)
A	40	41.32	1567	1.18E-8
B	65	43.58	2058	9.862E-9
C	80	44.62	2293	6.318E-9
D	130	39.2	4935	4.314E-9

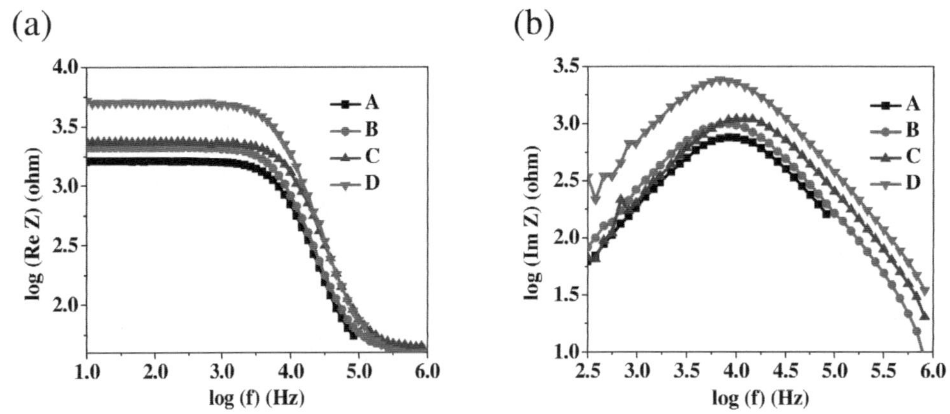

Figure 3. Frequency-dependent (a) real and (b) imaginary parts of the impedance of the PLEDs at 0 V.

Figures 3(a) and 3(b) show the frequency dependent real and imaginary parts of the impedance of devices with different EML thickness. Both real and imaginary parts of the impedance are changed with the EML thickness of the devices. The real and imaginary part of the impedance of device D increases significantly and can be attributed to the high parallel resistance and low parallel capacitance of the emissive layer due to its large thickness and high roughness which can be due to the less solubility of the polymer at high concentration.

The values of the contact resistance (R_S), the parallel resistance (R_P), and the parallel capacitances (C_P) extracted from the frequency-dependent real and imaginary parts of the impedance of the devices with different EML thickness are summarized in Table 1.

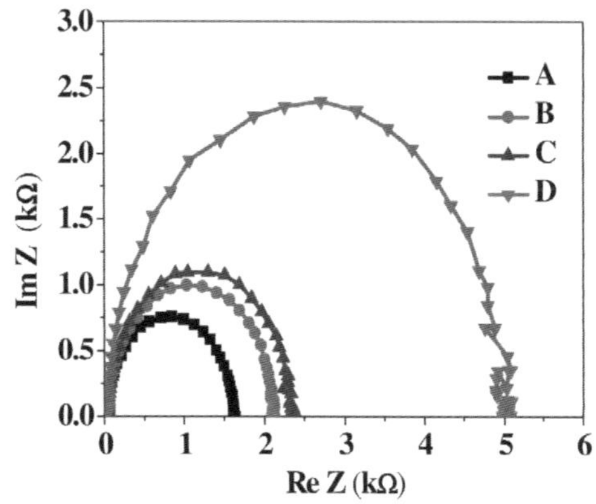

Figure 4. Impedance Cole-Cole plots for the devices with different emissive layer thickness

Figure 4 shows the impedance Cole-Cole plot for the devices with different EML thickness. The horizontal and the vertical axes represent the real (Z') and the imaginary (Z") parts of the impedance of the devices, respectively. All the devices have a very low series resistance suggesting efficient carrier injection from electrode as well as good interface charge transfer. However, as the thickness increases, the shunt resistance of the devices increases. As a result the impedance increases thereby increasing the driving voltage of the devices.

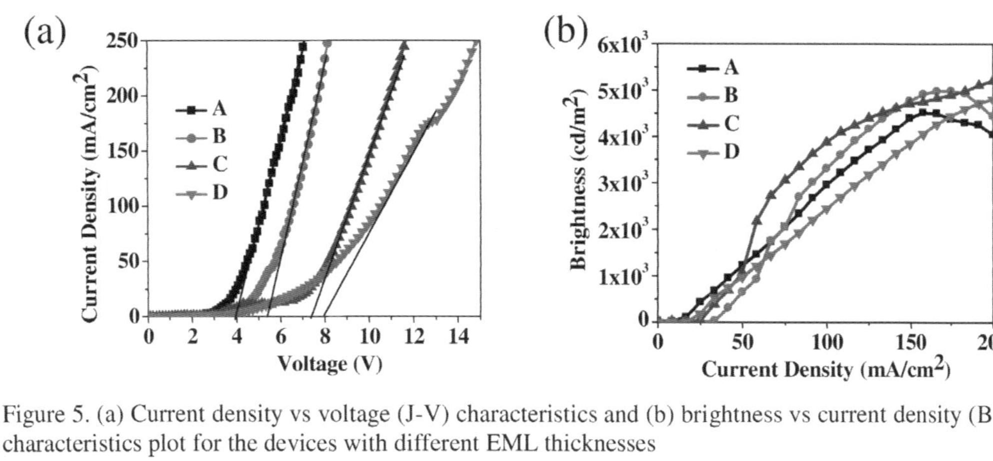

Figure 5. (a) Current density vs voltage (J-V) characteristics and (b) brightness vs current density (B-J) characteristics plot for the devices with different EML thicknesses

Figure 5(a) and 5(b) shows the current density vs voltage (J-V) characteristics and brightness vs current density (B-J) characteristics plot for the devices with different EML thicknesses. The driving voltage of the devices is found to increase with the increase in the EML thickness. However, in case of brightness vs current density curve, it has been noticed that the brightness is maximum for the device C having an EML thickness of approximately 80 nm. The key device properties of all the devices are listed in table 2. The turn on voltages of the PLEDs were calculated by extrapolating the straight portion of the J-V curves as shown in figure 5 (a) and is the voltage at which the device starts emitting light. It is observed that the turn-on-voltage varies from ~4 to 7.9 V when the EML thickness increases from 40 to 130 nm.

Table 2. Key device properties of the devices

Device Name	EML Thickness (nm)	Turn on Voltage (V)	Maximum Brightness (cd/m^2)
A	40	3.95	4516
B	65	5.40	5019
C	80	7.35	5225
D	130	7.93	4791

4. Conclusion

In conclusion, we have demonstrated the influence of EML thickness on the electrical characteristics of polymer light emitting diode (PLED). The impedance spectroscopy analysis showed that the devices with different EML thickness had different values of parallel resistance (R_P) and the parallel capacitance (C_P). The impedance of the devices is found to increase with increasing EML thickness resulting in an increase in the driving voltage. The device with an emissive layer thickness of 80nm, spin coated from a solution of concentration 15 mg/mL is found to give the best device performance with a maximum brightness value of 5226 cd/m^2.

Acknowledgments
The authors wish to thanks to the Department of Science and Technology (DST), India, (DST/TSG/PT/2009/11, DST/TSG/PT/2009/23), DST–Max Planck Society, Germany (No. IGSTC/MPG/PG(PKI)/2011A/48) and Department of Electronics & Information Technology, (Deity) No. 5(9)/2012-NANO (Vol. II) for the instrumental facilities. The author also wants to thanks IIT Guwahati for the financial support.

References

[1] Burroughes J H, Bradley D D C, Brown A R, Marks R N, Mackay K, Friend R H, Burns P L and Holmes A B 1990 *Nature* **347** 539
[2] Braun D and Heeger A J 1991 *Appl. Phys. Lett.* **58** 1982
[3] Gustafsson G, Cao Y, Treacy G M, Klavetter F, Colaneri N and Heeger A J 1992 *Nature* **357** 477
[4] Duggal A R, Heller C M, Shiang J J, Liu J, Lewis L N 2007 *J. Disp. Technol* **3**, 184
[5] Reineke S, Thomschke M, Lussem B and Leo K 2013 *Rev. Mod. Phys.* **85** 1245
[6] Carbas B B, Asil D, Friend R H and Onal A M 2013 *Org. Electron.* **15**, 500
[7] Sax S, Penkalla N R, Neuhold A, Schuh S, Zojer E, List E J W and Mullen K 2010 *Adv. Mater* **22**, 2087
[8] Neher D 2001 *Macromol. Rapid Commun.* **22** 1365
[9] Redecker M,. Bradley D D C, Inbasekaran M and Woo E P 1998 *Appl. Phys. Lett.* **73** 1565
[10] Huang J S, Li G, Wu E, Xu Q F and Yang Y 2006 *Adv. Mater.* **18** 114
[11] Gopikrishna P, Das D and Iyer P K 2015 *J. Mater. Chem. C.* **3** 9318
Das D, Gopikrishna P, Singh A, Dey A and Iyer P K 2016 *Phys. Chem. Chem. Phys.* **18** 7389

High Performance, Low Operating Voltage n-Type Organic Field Effect Transistor Based on Inorganic-Organic Bilayer Dielectric System

A Dey[1], A Singh[1], A Kalita[1], D Das[1] and P K Iyer[1, 2, *]

[1]Centre for Nanotechnology, Indian Institute of Technology Guwahati, Guwahati 781039, Assam, India
[2]Department of Chemistry, Indian Institute of Technology Guwahati, Guwahati 781039, Assam, India

*pki@iitg.ernet.in

Abstract. The performance of organic field-effect transistors (OFETs) fabricated utilizing vacuum deposited n-type conjugated molecule N,N˜-Dioctadecyl-1,4,5,8-naphthalenetetracarboxylic diimide (NDIOD2) were investigated using single and bilayer dielectric system over a low-cost glass substrate. Single layer device structure consists of Poly (vinyl alcohol) (PVA) as the dielectric material whereas the bilayer systems contain two different device configuration namely aluminum oxide/ Poly (vinyl alcohol) (Al_2O_3/PVA) and aluminum oxide/Poly (methyl methacrylate) (Al_2O_3/PMMA) in order to reduce the operating voltage and improve the device performance. It was observed that the devices with Al_2O_3/PMMA bilayer dielectric system and top contact aluminum electrodes exhibit excellent n-channel behaviour under vacuum compared to the other two structures with electron mobility value of 0.32 $cm^2/V\,s$, threshold voltages ~1.8 V and current on/off ratio ~10^4, operating under a very low voltage (6 V). These devices demonstrate highly stable electrical behaviour under multiple scans and low threshold voltage instability in vacuum condition even after 7 days than the Al_2O_3/PVA device structure. This low operating voltage, high performance OTFT device with bilayer dielectric system is expected to have diverse applications in the next generation of OTFT technologies.

1. Introduction

Organic field-effect transistors (OFETs) are the basic elements of organic electronic and optoelectronic applications. Due to their low-cost, large-area fabrication possibility, in near future OFET technologies are expected to build a new alternative market compared to conventional inorganic technologies in the mainstream electronics. Generally, OFETs are operated within large operating voltage range, which is one of the major limitations in finding their application in futuristic electronics [1-3]. Hence, an ideal OFET should have low operational voltage, high electrical stability and lifetime for real-life application, such as radio frequency identification (RFID) tag, sensor, electro-optical switch etc. It has already been reported that in order to achieve excellent device performance with low operating voltage, high-k inorganic dielectrics have been traditionally used [4, 5]. The popular choice has always been Al_2O_3 because of its high insulating property and low-cost deposition technique [6, 7]. However, the single Al_2O_3 film shows huge leakage current because of the inferior film forming quality. In contrast, Poly(vinyl alcohol) (PVA) and Poly(methyl methacrylate) (PMMA), the polymer dielectric materials, have emerged as the viable choice since they can be processed through solution processing methods, thereby reducing the overall cost of device fabrication. However, PVA-based OFETs often suffer from relatively low stability and large drain current (I_{DS}) hysteresis, which could be possibly associated due to the presence of -OH groups on the PVA surface [8]. On the other hand, due to the remarkable film forming ability and non-interacting nature of PMMA, it can be used to achieve high quality films in the organic-organic interface. However, these dielectric materials possess very low capacitance value, requiring very high threshold and operating voltages for device operation. Considering the advantages of polymer dielectric for better interfacial effect and high-k inorganic dielectric for low voltage operation, the concept of hybrid bilayer dielectric was proposed and introduced in OFETs.

In this study, we reported the fabrication and characterization of vacuum deposited NDIOD2 molecule containing 1, 4, 5, 8-Naphthalene diimides (NDI) substituted with octadecylamine (OD) chains, on Al_2O_3/PMMA bilayer dielectric system to achieve incredibly low threshold and operating voltages. The result was compared with our previous report containing single layer PVA gate dielectric on the glass substrate [9]. Al_2O_3 was used as a high dielectric constant material, whereas the role of PMMA was to provide a smoother interface to the organic semiconductor film. Devices having top contact aluminum electrode architectures exhibit excellent n-channel behavior with threshold voltage as low as 1.8 V. The ON/OFF current ratios of the devices were observed to be of the order of 10^4 up to an operating voltage of 6 V. Moreover, the devices exhibit remarkable stability under repeated scans and low threshold even after prolonged storage compared to Al_2O_3/PVA dielectric system. Thus, without compromising the carrier mobility (0.32 $cm^2/V\ s$) low-cost and environmentally friendly n-channel OTFT can be accomplished with drastic reduction in the operating voltage to 6 V from 50 V, by utilizing a combination of inorganic-organic (Al_2O_3/PMMA) bilayer dielectric system. This high performance OTFT device with 6 V operating voltage, with bilayer dielectric system is expected to have diverse applications in next generation of OTFT based technologies.

2. Experimental Section

2.1. *Chemicals and solvents*

1, 4, 5, 8-Naphthaleneteracarboxylic dianhydride, octadecylamine, quinoline, zinc acetate, PMMA (MW=550000 g/mol), and aluminum wire (99.999% purity) were used as received from Sigma Aldrich. PVA (MW= 1, 15,000 g/mol) and anisole were purchased from Loba Chemie (99% purity) and used as received. Microscope glass slides (thickness 1-1.2mm) purchased from Jain Scientific Glass Works, India, was used as the device substrate without any surface modification. NDIOD2 molecule was synthesized following the procedure reported previously [9].

2.2. *Characterization details*

The thickness of the thin films of NDIOD2, deposited by thermal evaporation technique, was measured by Veeco Dektak 150 Surface Profilometer. All AFM images were recorded by Agilent 5500-STM instrument under non-contact mode. Gaussian 03 software was used to perform the DFT simulation of the material. CH Instrument was used to perform the electrochemical analysis of NDIOD2. Finally, all the electrical properties were recorded by Keithley 4200 semiconductor characterization system (SCS).

3. Result and Discussion

3.1. *Band gap analysis*

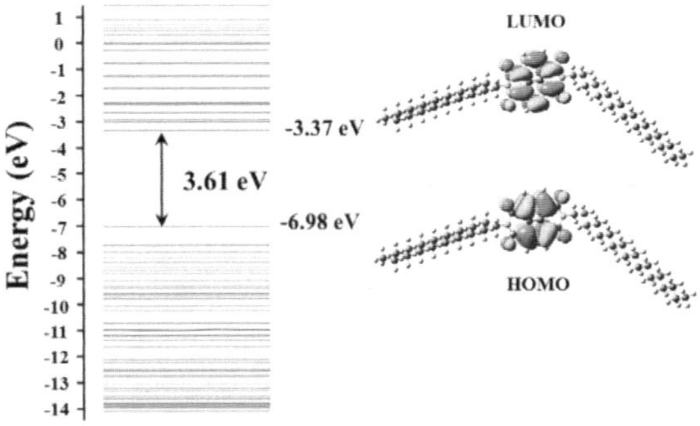

Figure 1. Illustration of energy level diagram of NDIOD2 molecule as estimated from DFT calculation.

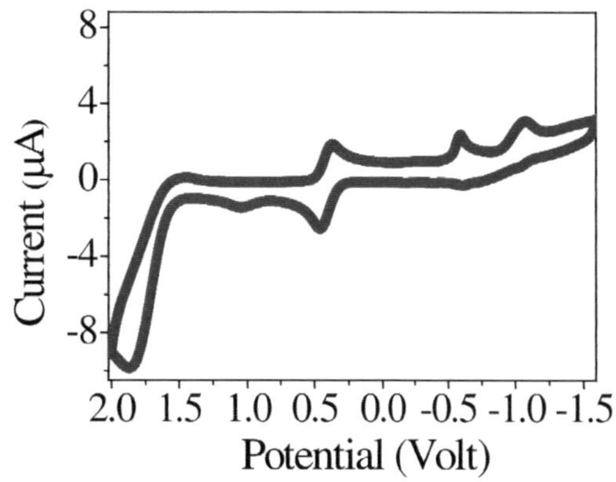

Figure 2. Cyclic voltametric curve of NDIOD2 molecule thin film with ferrocene as the reference.

The geometrically optimized chemical structure of NDIOD2 obtained from DFT calculation by considering B3LYP/6-31G(d) basis set along with its energy level diagram and the cyclic voltametric curve of NDIOD2 molecule are represented in Figure 1 and 2 respectively. The HOMO and LUMO energy values of NDIOD2 molecule, estimated by electrochemical analysis using onset calculation of Equations (1) and (2) are given below-[10]

$$E_{HOMO} = -(E_{OX,onset} + 4.8)eV \quad (1)$$

$$E_{LUMO} = -(E_{RED,onset} + 4.8)eV \quad (2)$$

where, -4.8 eV is the absolute HOMO level of ferrocene, acting as the internal reference. The estimated HOMO and LUMO energy levels and the band gap of the material were summarized in Table 1.

Table 1. Summary of Band Gap Calculation Data for NDIOD2[a].

$E_{HOMO}^{(CV)}$	$E_{LUMO}^{(CV)}$	$E_g^{(CV)\,b}$	$E_g^{(Th)\,c}$
-6.414	-3.394	3.02	3.61

[a] Energies are in eV.
[b] Electrochemical band gap.
[c] Theoretical band gap was calculated by using DFT calculation

3.2. *Device fabrication method on PVA single layer dielectric configuration*

OTFTs based on NDIOD2 molecule were fabricated following the procedure reported in our previous study [9]. Glass slides were used as substrates onto which an aluminum gate (thickness 100 nm) was deposited by thermal evaporation method through a shadow mask. Further, a solution of PVA dielectric in de-ionized water (10 wt. %) was spin-coated on top of the aluminum film and dried for 1 h at 100 °C in a vacuum oven. The thickness of the PVA film was approximately 1μm, measured by surface profilometer, which showed capacitance of ~8.854 nF.cm^{-2}. Following this, a 60 nm thick (±10 nm) NDIOD2 thin film was thermally deposited at room temperature, at a pressure of 10^{-6} mbar. Finally, square shaped aluminum source and drain contacts having thickness ~100 nm were thermally evaporated to characterize the electrical parameters of the OFETs (using a Keithley 4200 semiconductor characterization system (SCS)) under vacuum. The schematic structure of the device and the AFM images of PVA dielectric as well as the NDIOD2 molecule on top of PVA dielectric layer using glass substrate are shown below in Figure 3 and 4 respectively.

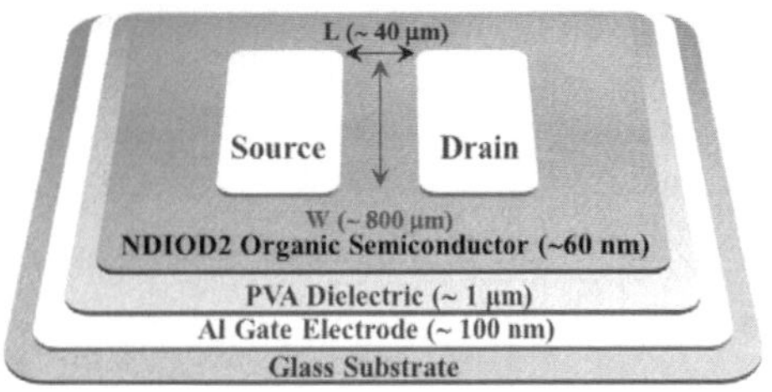

Figure 3. Schematic of the device structure with single layer PVA dielectric system.

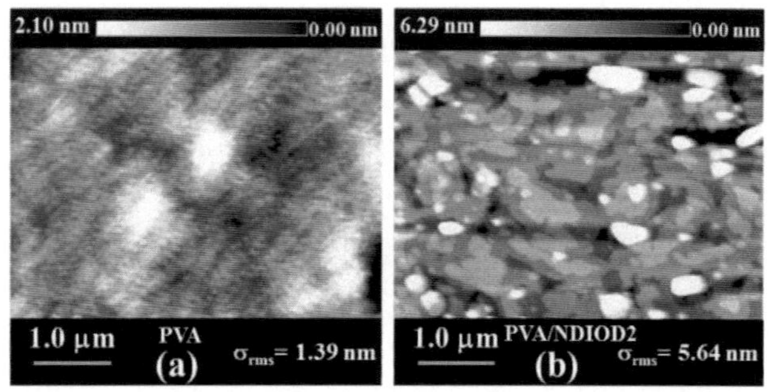

Figure 4. AFM image of (a) PVA dielectric thin film (~1 μm) and (b) PVA/NDIOD2 semiconductor on the glass substrate.

3.3. *Device fabrication method on Al_2O_3/PVA and Al_2O_3/PMMA bilayer dielectrics configurations*

Anodization is a very good, effective and solution based technique to grow metal oxide films with nanometer control. The high capacitance, pinhole-free Al_2O_3 gate insulators can be fabricated by electrochemical oxidation (or anodization) of the gate metal on desired substrates, such as, glass and flexible transparent sheets etc. In addition, this process can yield a high-quality metal-oxide insulator at room temperature with very low cost and less time.

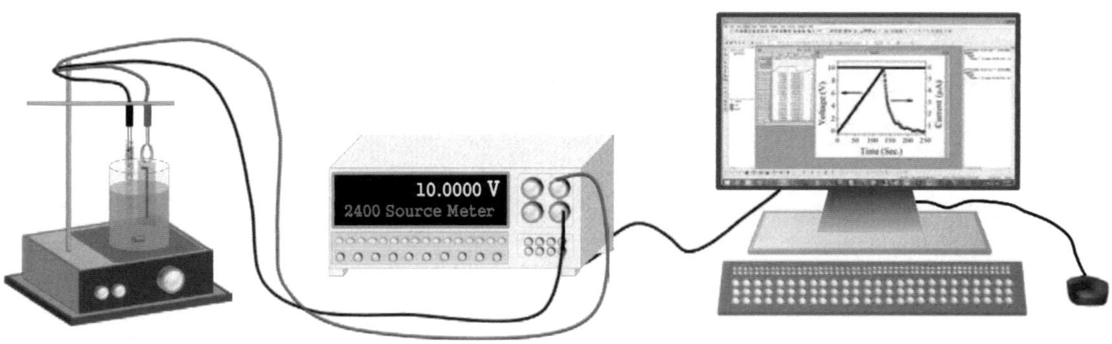

Figure 5. Anodic oxidation setup for aluminum.

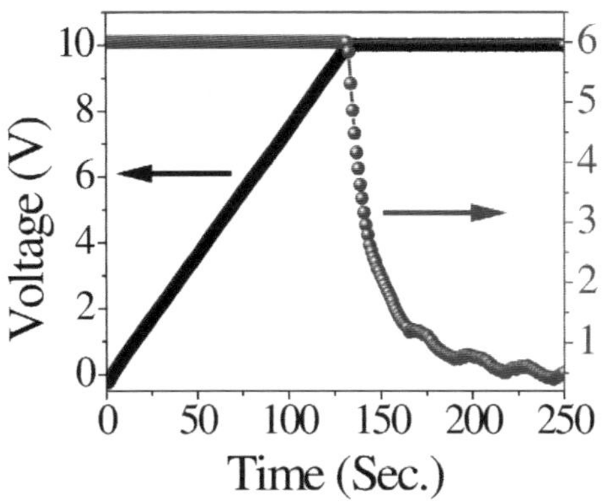

Figure 6. Anodic oxidation curve for aluminum.

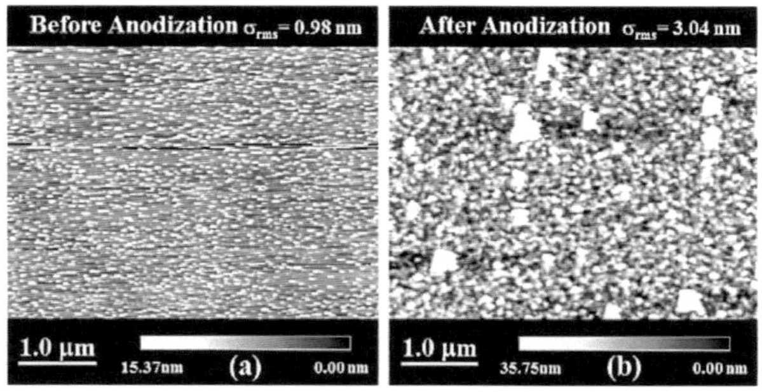

Figure 7. AFM image of aluminum thin film (a) before and (b) after the anodization.

In this method we used cleaned glass slides (size: 15 mm x 25 mm) to serve as device substrates, onto which >200 nm thick aluminum gate with dimensions 1 mm x 20 mm was deposited by thermal evaporation method through a shadow mask. The film was then anodized with a constant current density of 0.06 mA cm^{-2} and a voltage of 10 V in a 0.001 M citric acid monohydrate electrolyte solution at 25°C (room temperature) using a square-shaped platinum mesh as counter electrode to form a ~13 nm thick Al_2O_3 layer over the aluminum film gate electrode. Figures 5, 6 and 7 demonstrate the anodization setup which was used for growing Al_2O_3 film, the typical anodization graph and the AFM images of aluminum film before and after anodic oxidation.

After iodization, to further reduce the surface roughness, a 100 nm thick PVA and PMMA thin film was separately spin coated on the two device configurations, to form Al_2O_3/PVA and Al_2O_3/PMMA bilayer dielectric systems, and dried for 1 h at 120°C under nitrogen atmosphere. Both the organic dielectric solution was prepared by dissolving 30 mg/ml of PVA in de-ionized water and 30 mg/ml of PMMA in anisole. The capacitance of the Al_2O_3/PVA and Al_2O_3/PMMA bilayer dielectric systems were observed as ~6 nF.cm^{-2} and ~27 nF.cm^{-2} respectively, out of which Al_2O_3/PMMA is much higher than the single layer device structure. Followed by this, 60 nm (± 10 nm) the NDIOD2 active material was deposited by thermal deposition method under a base pressure of 10^{-6} mbar. Further, aluminum source-drain electrodes were thermally evaporated, at room temperature, up to a thickness of 100 nm to calculate three terminal properties of the NDIOD2 material. The schematic of the device structures and the AFM images of Al_2O_3/PVA and Al_2O_3/PMMA bilayer dielectrics and NDIOD2 molecule on top of these layers using glass substrates are shown below in Figures 8 - 11 respectively.

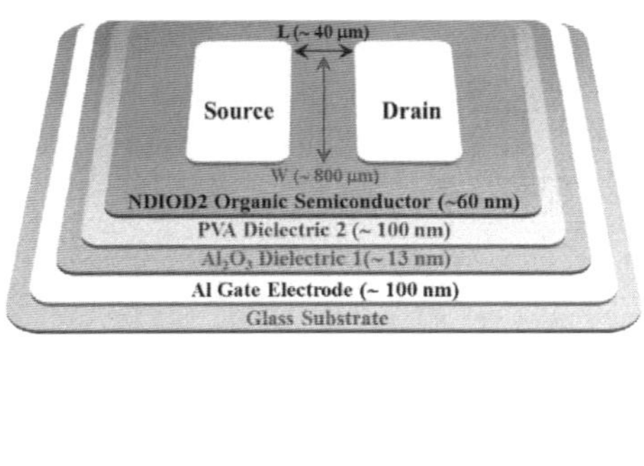

Figure 8. Schematic of the device structure with Al_2O_3/PVA bilayer dielectrics system.

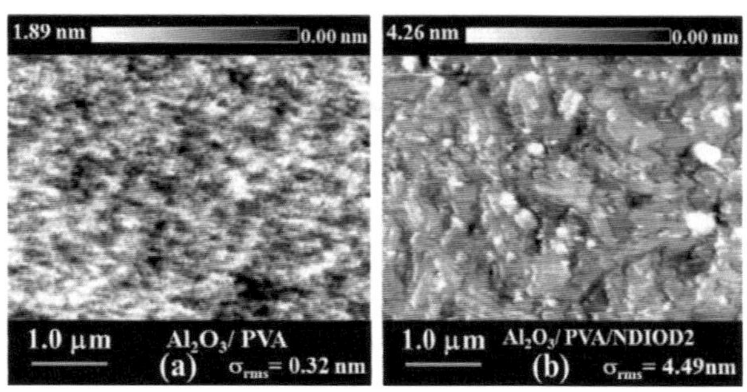

Figure 9. AFM image of (a) PVA dielectric thin film (~100 nm) on Al_2O_3 and (b) Al_2O_3/PVA/NDIOD2 semiconductor on the glass substrate.

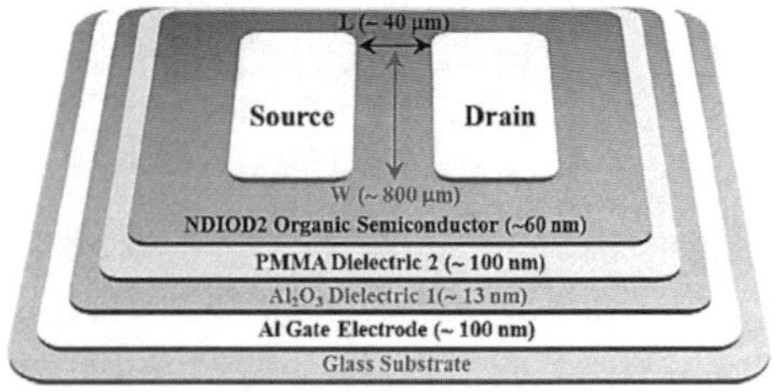

Figure 10. Schematic of the device structure with Al_2O_3/PMMA bilayer dielectrics system.

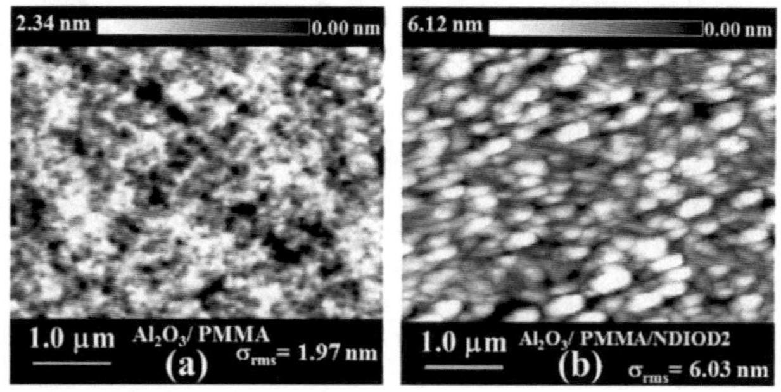

Figure 11. AFM image of (a) PMMA dielectric thin film (~100 nm) on Al_2O_3 and (b) Al_2O_3/ PMMA/NDIOD2 semiconductor on the glass substrate.

3.4. *Device Characterisation*

The OFET mobility value of the NDIOD2 molecule was estimated from the saturated region by using the following Equation (3).

$$I_{DS} = \frac{\mu W C_i}{2L}\left(V_G - V_{Th}\right)^2 \quad (3)$$

where I_{DS}, µ and C_i, represent the drain current, the field effect electron mobility and the capacitance per unit area of the bilayer gate dielectric. V_{Th} and V_G signify the threshold voltage and gate voltage respectively. Figures 12, 13 and 14 demonstrate the typical drain and transfer characteristics curve of NDIOD2 molecule on single layer PVA, bilayer Al_2O_3/PVA and Al_2O_3/PMMA dielectric system respectively. The calculated OFET parameters of the molecule are listed in Table 2.

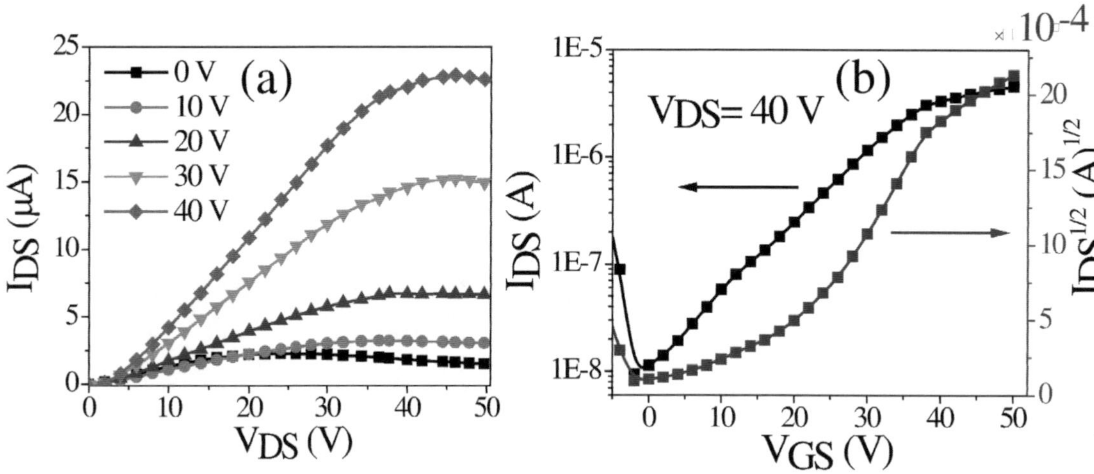

Figure 12. (a) Drain and (b) transfer characteristics curve of NDIOD2 molecule in the PVA single layer dielectric system.

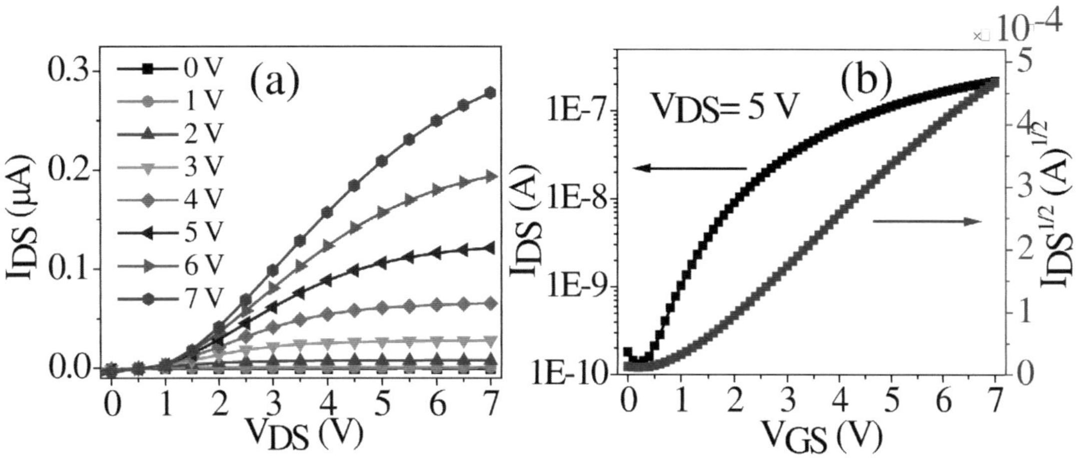

Figure 13. (a) Drain and (b) transfer characteristics curve of NDIOD2 molecule in the Al_2O_3/PVA bilayer dielectrics system.

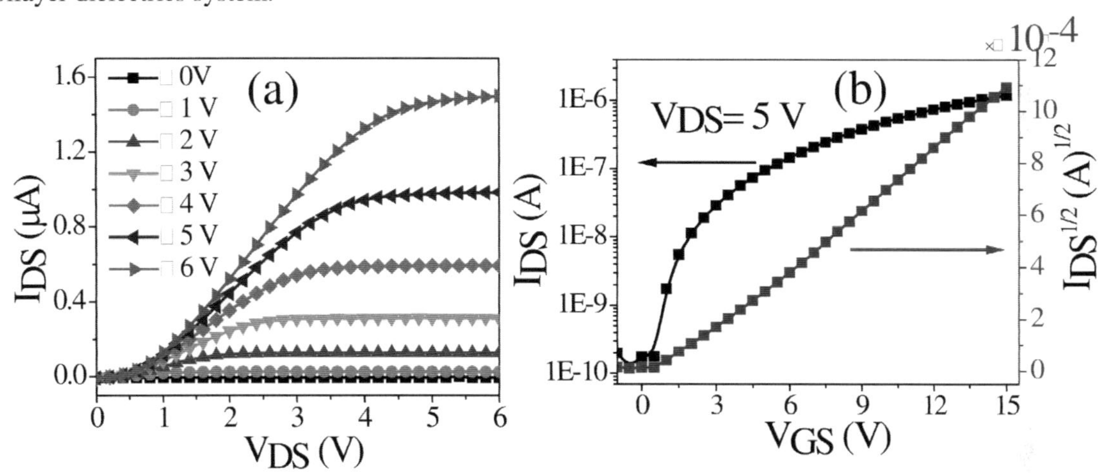

Figure 14. (a) Drain and (b) transfer characteristics curve of NDIOD2 molecule in the Al_2O_3/PMMA bilayer dielectrics system.

Table 2. Comparison of OFETs measurement for NDIOD2 molecule between single and bilayer dielectrics system

Dielectric	Thickness	Capacitance(C_i) (nF/cm^2)	Mobility(μ) (cm^2/V.s)	V_{Th}(V)	I_{on}/I_{off}
PVA	1μm	8.854	0.68	16.24	10^2
Al_2O_3/PVA	10nm/100nm	6	0.13	0.99	10^3
Al_2O_3/PMMA	10nm/100nm	27	0.32	1.8	10^4

The NDIOD2 molecules in OFET devices, sublimed at room temperatures over bilayer Al_2O_3/PMMA gate dielectric, exhibited excellent average electron mobility as 0.32 cm^2/Vs under vacuum with a threshold voltage of 1.8 V. The ON/OFF current ratio of the devices were observed to be of the order of 10^4 up to an operating voltage of only 6 V. However, the device with Al_2O_3/PVA showed low threshold voltage and low electron mobility likely due to the presence of -OH group in the PVA dielectric which creates charge traps at the dielectric-semiconductor interface. These traps also reduce the overall stability and degrade the devices very fast (~1 day). Thus, the Al_2O_3/PMMA bilayer dielectric system demonstrates drastic improvement in the device performance due to better capacitive coupling between the gate and the channel through the dielectric layer to enhance field-effect carrier mobility and the threshold voltage compared to the PVA dielectric system.

4. Conclusion

In conclusion, we described the electrical performances of vacuum-deposited thin film based n-channel organic field effect transistors with alkyl chain-substituted NDIOD2 molecule. The devices were studied in a top contact bottom gate configuration using aluminum source/drain electrodes with PVA, Al_2O_3/PVA and Al_2O_3/PMMA as single and bilayer gate dielectric systems respectively. The devices fabricated using NDIOD2 with Al_2O_3/PMMA gate dielectric and aluminum source-drain metal electrodes exhibited excellent n-channel behavior with average field-effect electron mobility of 0.32 cm^2/V s and threshold voltage of 1.8 V for channel length, L=40 μm and channel width, W=800 μm. This OFET device showed ON/OFF current ratio of 10^4 which is much higher than PVA dielectric based devices. Also the Al_2O_3/PMMA devices demonstrated highly stable electrical behavior with no hysteresis when tested at low vacuum conditions under several scans, even after 7 days. These low cost, high performance devices are expected to open newer avenues in the field of electronic devices, energy efficient sensors and stable switching devices.

Acknowledgments

The authors thanks the Department of Science and Technology (DST), India, (DST/TSG/PT/2009/11, DST/TSG/PT/2009/23), DST–Max Planck Society, Germany (No. IGSTC/MPG/PG (PKI)/2011A/48) and Department of Electronics & Information Technology, (Deity) No. 5(9)/2012-NANO (Vol. II) for financial support. The Central Instruments Facility, IIT Guwahati is acknowledged for providing sophisticated instrument facilities.

References

[1] Ong B S, Wu Y, Liu P and Gardner S 2004 *J. Am. Chem. Soc.* **126** 3378.
[2] Li Y, Wu Y, Liu P, Birau M, Pan H and Ong B S 2006 *Adv. Mater.* **18** 3029.
[3] Noh Y -Y, Kim D-Y, Yoshida Y, Yase K, Jung B –J, Lim E and Shim H –K 2005 *Appl. Phys. Lett.* **86** 043501.
[4] Baeg K -J, Noh Y -Y, Ghim J, Kang S –J, Lee H and Kim D -Y 2006 *Adv. Mater.* **18** 3179.
[5] Noh Y-Y, Zhao N, Caironi M and Sirringhaus H 2007 *Nat. Nanotechnol.* **2** 784.
[6] Iba S, Sekitani T, Kato Y, Someya T, Kawaguchi H, Takamiya M, Sakurai T and Takagi S 2005 *Appl. Phys. Lett.* **87** 023509.
[7] Koo J B, Ku C H, Lim S C, Kim S H and Lee J H 2007 *Appl. Phys. Lett.* **90** 133503.
[8] Kim J, Lee J, Kim J, Ju B, Kim J, Lee Y and Oh M 2004 *Appl. Phys. Lett.* **85** 6368.
[9] Dey A, Kalita A and Iyer P K 2014 *ACS Appl. Mater. Interfaces* **6** 12295.
[10] Jones B A, Facchetti A, Wasielewski M R, Marks and T J 2007 *J. Am. Chem. Soc.* **129** 15259.

Improved characteristics of polymer light-emitting diode based on poly(alkylfluorene) with functional layer fabricated by contact printing method

H Kajii, K Mizuno and Y Ohmori

Graduate School of Engineering, Osaka University, 2-1 Yamada-oka, Suita, Osaka 565-0871, Japan

E-mail: kajii@oled.eei.eng.osaka-u.ac.jp

Abstract. In the organic light-emitting diode, the multilayer structure is useful to achieve the effective carrier injection and transport, and improve the emission efficiency. The characteristics of the polymer light-emitting device with a functional layer by contact printing method are investigated. Hole transport material, the guest and electron transport material are mixed at each layer with the same host. Poly(9,9-dioctylfluorene) (F8) and Tris[1-phenylisoquinolinato-C2,N]iridium(III) ($Ir(piq)_3$) are used as the host and guest materials. We focused on the effects of a functional layer with 1,3,5-Tris (1-phenyl-1H-benzimidazol-2-yl)benzene(TPBi) as the electron transport material doped in F8. For a non-heterostructure device with F8:TPBi functional layer, electrons are injected from the cathode interface to TPBi and transported through F8. TPBi acts as the effective hole blocking and electron transport layer as TPBi is mixed in F8 at the cathode side. Non-heterostructure device exhibits improved emission efficiency and lower driving voltage.

1. Introduction

In organic semiconductors, π-conjugated polymers, being quasi-one-dimensional macromolecular electronic systems, offer a number of unique properties. Blue polymer light-emitting diodes utilizing poly(alkylfluorene) have been demonstrated for the first time [1]. Fluorene-type polymers have emerged as an important class of conducting polymers due to their efficient emission, high stabilities and relatively high mobilities [2-4]. Fluorene-type polymers also have the potential for full-color emission via energy transfer to longer wavelength emitters in blends with other fluorescent and phosphorescent dyes [5].

Employing phosphorescent materials yields high efficiencies because breaking the spin conservation rule allows both singlet and triplet excitons to contribute to emission. A specific class of organic light-emitting diodes (OLEDs) has achieved high luminescence efficiency using the phosphorescent emission from a triplet state of Ir complexes [6,7]. Employing solution process of organic materials is useful for OLEDs fabrication, and allows for the possibility of low cost mass production and for making high-sized screens.

In the OLEDs, the multilayer structure is useful to achieve the effective carrier injection and transport, and to improve emission efficiency. We investigated the characteristics of the polymer light-emitting diode (PLED) based on poly(alkylfluorene) with a functional layer by contact printing method.

2. Materials and Methods

The glass substrate was degreased with solvents and cleaned in a UV ozone chamber. To reduce the possibility of electrical shorts within the PLEDs, a 40-nm-thick poly(ethylenedioxythiophene): poly(styrene sulfonic acid) (PEDOT:PSS) hole-injection layer was spin-coated on an indium tin oxide (ITO)-coated glass substrate. Poly(9,9-dioctylfluorene)(F8) and tris[1-phenyliso quinolinato-C2,N]iridium(III) (Ir(piq)$_3$) were used as the host and the guest materials. The doping concentration of Ir(piq)$_3$ was fixed at 3 wt.% relative to F8. Hole transport material, guest and electron transport material were mixed at each layer with the same host. To improve carrier balance, a hole transport material of 4,4',4"-Tri(9-carbazoyl)triphenylamine (TCTA) are blended into an emissive layer. 1,3,5-Tris (1-phenyl-1H-benzimidazol-2-yl)benzene (TPBi) was used as the electron transporting material. The emissive layers (EMLs) were fabricated by spin-coating method into 50 or 70 nm-thick layers on PEDOT:PSS layer. The concentration of the EML of F8:TCTA:TPBi:Ir(piq)$_3$ was 100:50:40:3 wt.%. A TPBi layer was fabricated by organic molecular beam deposition on the EML to form the device at a background pressure of about 10^{-5} Pa.The cathode consisting of CsF (2 nm)/Mg:Ag/Ag or MoOx (5 nm)/Ag and Al anode were deposited in vacuum at a chamber base pressure of about 10^{-4} Pa. The active area was 4 mm^2.

The absolute photoluminescence quantum yields (PLQYs) were measured using an absolute PL quantum yield measurement system (HAMAMATSU, Quantaurus-QY). The film for the PLQY measurement was formed on quartz glass substrate. The electroluminescence (EL) spectra were measured using a photonic multichannel spectral analyzer (Hamamatsu Photonics, PMA-11). The current density-voltage-luminance (*J-V-L*) characteristics were obtained using a digital sourcemeter (Keithley2400), and a luminance meter (Minolta LS-100). Impedance spectroscopy measurements were carried out using a Solatron 1260 impedance analyzer with a 1296 dielectric interface. The typical AC oscillation amplitude was 100 mV.

3. Results and Discussion

A multilayer approach to fabricating active layers helps to improve PLED performance. We focused on the improved characteristics of a non-heterostructure PLED with a functional layer based on TPBi doped in F8. The device structures are as follows:

device A: ITO/PEDOT:PSS/F8:TCTA:Ir(piq)$_3$ (70 nm)/CsF/Mg:Ag/Ag,
device B: ITO/PEDOT:PSS/F8: TCTA:TPBi:Ir(piq)$_3$ (70 nm)/CsF/Mg:Ag/Ag,
device C: ITO/PEDOT:PSS/F8:TCTA: Ir(piq)$_3$ (50 nm)/F8:TPBi (20 nm)/CsF/Mg:Ag/Ag,
device D: ITO/PEDOT:PSS/F8:TCTA: Ir(piq)$_3$ (50 nm)/TPBi (20 nm)/CsF/Mg:Ag/Ag.

For device C, a F8:TPBi layer, deposited on a poly(dimethysiloxane) stamp, was transferred to the EML using contact printing [8]. For comparison, device D with TPBi as a hole-blocking layer was fabricated.

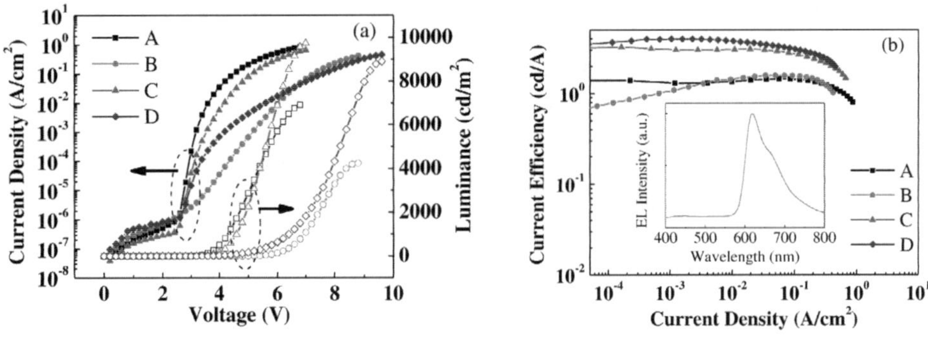

Figrure 1. (a) *J-V-L* and (b) current efficiency–current density characteristics of devices A, B, C and D. The inset of Fig. 2b shows typical EL spectrum of device C.

PLQY of an EML is one of the important factors to achieve the high efficiency. The energy transfer between other host materials and $Ir(piq)_3$ mainly occurred via the triplet state of host materials and the metal ligand charge transfer (MLCT) state of $Ir(piq)_3$. The excited energy state also transferred from the singlet state of host materials to the singlet MLCT state, and then transferred to the triplet (MLCT) state via intersystem crossing. PLQYs of F8:TCTA:$Ir(piq)_3$ and F8:TCTA:TPBi:$Ir(piq)_3$ films were estimated to be 40 and 41 %. PLQY of EML in device B was almost the same as those in devices A, C and D. All devices exhibited red emission based on $Ir(piq)_3$ as shown in the inset of Fig. 1b.

Figure 1 shows the current density-luminance-applied voltage (*J-V-L*) and current efficiency – current density characteristics of devices A, B, C and D. Devices C and D showed increased luminance and current efficiency, as compared to devices A and B. For device C, the maximum luminance of approximately 10,000 cd/m^2 was obtained. Devices A and C showed lower driving voltage than devices B and D.

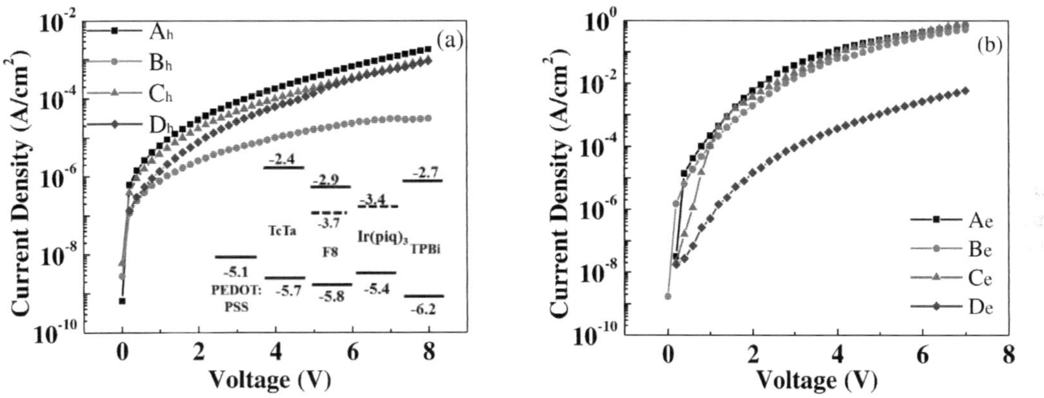

Figure 2. *J-V* characteristics of (a) hole-only devices A_h, B_h, C_h and D_h, and (b) electron-only devices A_e, B_e, C_e and D_e. The inset of Fig. 2a shows schematic energy diagram of organic materials used in this study.

We fabricated four types of hole-only devices A_h, B_h, C_h and D_h with MoOx/Ag cathode instead of CsF/Mg:Ag/Ag to understand hole injection and transport. Hole current of device B was markedly lower than those of devices A_h, C_h and D_h, as shown in Fig. 2a. Hole currents of devices C_h and D_h were also slightly decreased, compared to device A. This result suggested that TPBi acts as the hole blocking as TPBi has the lowest unoccupied molecular orbital (LUMO) level lower than other materials as shown in the inset of Fig. 2a. Device B_h, which had a TPBi-mixed emissive layer, showed the worse hole injection and transport, which results in higher driving voltage for device B.

In order to estimate the effect of electron injection and transport, current density-voltage characteristics of four types of electron-only devices A_e, B_e, C_e and D_e with Al anode instead of ITO/PEDOT:PSS are given in Fig. 2b. Electron current of device D_e was markedly lower than those of devices A_e, B_e and C_e, which results in higher driving voltage for device D.

The fundamental properties of devices are mostly investigated under DC operation. The *J*–V characteristics represent the overall result of injection and transport of carriers through the organic layers and electrodes. From the view of carrier transport dynamics, impedance spectroscopy is an effective method to study transport dynamics in the devices [9-12]. To further understand the factors affecting the electrical properties due to TPBi, these behaviors of the electrical properties of the devices were analyzed using impedance spectroscopy measurements. Figure 3 shows the Bode plot of the imaginary part of the modulus in devices A, B and C. The peak frequencies of the Bode-plot show the reciprocal value of the relaxation times. For all devices, the relaxation times of the emission layer increased with increasing voltage. The peak shift to a higher frequency is related to the decrease in resistance of the organic layers. Compared to devices B and C, an additional peak for device A was observed at high frequencies. This result suggested that for device A, CsF induces transient accumulation of electrons at the F8/electrode interface. For the devices B and C with mixed F8 and TPBi, electrons are injected from the cathode interface to TPBi and transported through F8. Electron transport ability of F8 is better than that of TPBi. For device C, F8:TPBi layer acts as hole blocking and electron transport layer, which results in improved carrier balance.

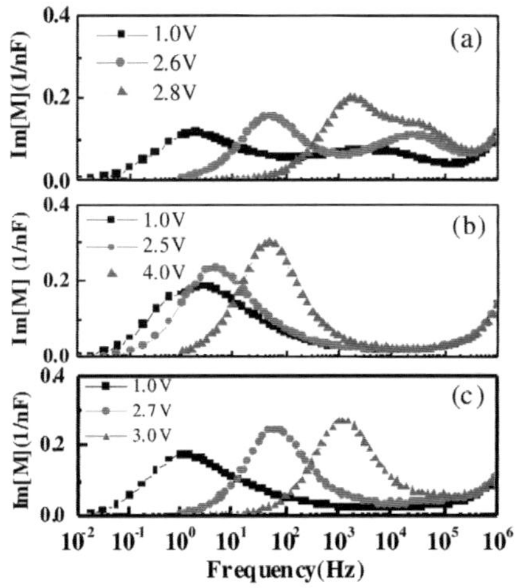

Figure 3. Bode-plot of the imaginary part of the modulus in (a) device A, (b) device B and (c) device C.

4. Conclusions

We investigated the characteristics of the polymer light-emitting devices with functional layers by contact printing method. We focused on the effects of the F8:TPBi functional layer. Non-heterostructure device showed the maximum luminance of approximately 10,000 cd/m^2. For the non-heterostructure device with a F8:TPBi functional layer, electrons are injected from the cathode interface to TPBi and transported through F8. TPBi acts as the effective hole blocking and electron transport layer as TPBi is mixed in F8 at the cathode side. Non-heterostructure device exhibited improved emission efficiency and lower driving voltage owing to improved carrier balance and transport.

Acknowledgements

Part of this work was supported by Grant-in-Aid for Scientific Research, from the Ministry of Education, Culture, Sports, Science and Technology, Japan. This research was also partially supported financially by a Grant in Aid of Special Coordination Funds for Promoting Science and Technology.

References

[1] Ohmori Y, Uchida M, Muro K and Yoshino K 1991 *Jpn. J. Appl. Phys.* **30** L1941-L1943
[2] Kajii H, Koiwai K, Hirose Y and Ohmori Y 2010 *Org. Electron.* **11** 509-513
[3] Koiwai K, Kajii H and Ohmori Y 2011 *Synth. Met.* **161** 2107-2112
[4] Tanaka H, Kajii H and Ohmori Y 2015 *Synth. Met.* **203** 10-15
[5] Hino Y, Kajii H and Ohmori Y 2006 *Thin Solid Films* **499** 359-363
[6] Baldo M A, Lamansky S, Burrows P E, Thompson M E and Forrest S R 1999 *Appl. Phys. Lett.* **75** 4-6
[7] Adachi C, Baldo M A, Forrest S R and Thompson M E 2000 *Appl. Phys. Lett.* **77** 904-906
[8] Takakuwa A, Misaki M, Yoshida Y and Yase K 2009 *Thin Solid Films* **518** 555-558
[9] Meier M, Karg S and Riess W 1997 *J. Appl. Phys.* **82** 1961-1966
[10] Scherbel J, Nguyen P H, Paasch G, Brütting W and Schwoerer M 1998 *J. Appl. Phys.* **83** 5045-5055
[11] Okachi T, Nagase T, Kobayashi T and Naito H 2008 *Thin Solid Films* **517** 1327-1330
[12] Yamamoto T, Kajii H and Ohmori Y 2014 *Org. Electron.* **15** 1077-1082

Growth promotion effect of steelmaking slag on *Spirulina platensis*

R Nogami[1], L T Tam[2], H T L Anh[2], H T H Quynh[2], L T Thom[2], P V Nhat[2], N T H Thu[2], D D Hong[2] and M Wakisaka[1]

[1]Graduate School of Life Science and Systems Engineering, Kyushu Institute of Technology, 2-4 Hibikino, Wakamatsu, Kitakyushu, Fukuoka, 808-0196, Japan
[2]Institute of Biotechnology, Vietnam Academy of Science and Technology, 18 Hoang Quoc Viet Road, Cau Giay District, Hanoi, Vietnam

E-mail address: wakisaka@life.kyutech.ac.jp

Abstract. A growth promotion effect of steelmaking slag on *Spirulina platensis* M135 was investigated. The growth promotion effect was obtained that was 1.27 times greater than that obtained by the control by adding 500 mg L^{-1} of steelmaking slag and culturing for 60 days. The lipid content decreased in a concentration-dependent manner with steelmaking slag, whereas the carbohydrate content remained constant. The protein content of *S. platensis* M135 increased in a concentration-dependent manner with steelmaking slag when cultured at day 45. The superoxide dismutase activity of *S. platensis* M135 exhibited a decreasing trend in a time-dependent manner and an increasing trend in the control. The superoxide dismutase activity was lower than that of the control at day 1 but was higher at day 30. No genetic damage was observed up to 500 mg L^{-1} of steelmaking slag at 30 days of culture. Recovery from genetic damage was observed at 1,000 mg L^{-1} of steelmaking slag but not at higher concentrations.

1. Introduction

Global warming because of the increasing levels of greenhouse gases, including carbon dioxide, is a serious concern. The potential for carbon capture by biomass production has been attracting much attention; in particular, the higher productivity of microalgae than that of plant biomass has been of interest. *Spirulina* is one of the most popular edible microalgae. It is rich in protein and can grow in severe conditions such as high temperatures, salinity, and highly alkaline pH. Because of these specific culture conditions, *Spirulina* can avoid contamination of other microbes; thus, it can be easily commercialized. Screening for strains with higher growth rates and higher concentrations of targeted products is required for commercialization. Genetic modification is expected to be applicable to *Spirulina*, since the genetic structure of *Spirulina* (*Arthrospira*) *platensis* NIES-39 has been elucidated [1]. But application of genetic modified microorganism should be severely restricted. Much more simple approach is supplementation of growth enhancer to culture medium. *Spirulina* growth promotion is easily achieved by adding a carbon source such as glucose. Chen *et al.* reported a growth promoting effect of 2.38 times than that of the control by adding 2 g L^{-1} glucose and of 1.53 times that of the control by adding 4 g L^{-1} acetic acid [2]. Nevertheless, even though higher productivity is achieved, these approaches are costly. Therefore, more cost effective methods to produce biomass have been investigated such as the utilization of seawater or organic wastewater as a culture medium. Ak reported that cultivation costs could be reduced by 20% using wastewater [3].

Iron and steel slag is a by-product of iron and steel manufacturing processes and can be categorized into blast furnace slag and steelmaking slag (SMS). SMS consists of both converter slag that is generated by a converter and electric arc furnace slag that is generated during the electric arc furnace steelmaking process that uses steel-scrap as its raw material. In Japan, 15.627 million tons of SMS slag was annually produced in 2014, which was used in a wide range of applications such as cement, civil engineering works at ports, concrete aggregate, and ground improvement [4]. But SMS can be applicable as a fertilizer because it contains the plant nutrients such like CaO, SiO_2, MgO, and FeO. Fertilization effect of SMS for seaweed bed restoration has been actually proved in various Japan seacoasts [5-8].

The ecological impact of SMS and its component on marine organisms and on some seawater microalgae has been concerned. For example, ferric ethylenediaminetetraacetic acid (Fe-EDTA) effectively increased the growth of the macro algae *Sargassum horneri* [9]. Liu *et al.* reported a growth promoting effect of $FeCl_3$-EDTA on *Chlorella vulgaris* [10]. The effect of 10 mg L^{-1} iron nanoparticles (Total Fe concentration $<1.4 \times 10^{-5}$ M) on the growth and metabolic status of various marine microalgae such as *Pavlova lutheri*, *Isochrysis galbana*, and *Tetraselmis suecica* has been investigated [11]. Iron binding at the cell surface has been reported to be a critical step in the uptake mechanism of marine microalgae [12]. To maximize the iron fertilizer effect, the size of the SMS should be small to increase the surface area for elution. Normally the size of the SMS should be carefully determined or controlled to avoid sudden increase in alkalinity because of CaO reacting with water. But this disadvantage of SMS inducing high alkalinity in water is expected to be advantage for growth of *Spirulina*, specifically. Therefore, the effect of SMS on growth, and the metabolite and genetic status of *S. platensis* M135 was investigated in this study.

2. Materials and methods

2.1. Materials

The cyanobacteria *S. platensis* strain M135 was supplied by the Algal Biotechnology Department, Institute of Biotechnology, Vietnam Academy of Science and Technology, Hanoi, Vietnam. *S. platensis* M135 were cultured in 250 mL flasks with 100 mL of SOT medium with the following composition (mg L−1): $NaHCO_3$, 16,800; K_2HPO_4, 500; $NaNO_3$, 500; K_2SO_4, 1,000; NaCl, 1,000; $MgSO_4 \cdot 7H_2O$, 200; $CaCl_2 \cdot 2H_2O$, 40; $FeSO_4 \cdot 7H_2O$, 10; $Na_2EDTA \cdot 2H_2O$, 80; H_3BO_3, 2.86; $MnSO_4 \cdot 5H_2O$, 2.5; $ZnSO_4 \cdot 7H_2O$, 0.22; $CuSO_4 \cdot 5H_2O$, 0.08; $Na_2MoO_4 \cdot 2H_2O$, 0.02. *S. platensis* M135 was cultivated under the following conditions: light intensity of 1,500 Lux from a white fluorescent lamp with 12 h/12 h light/dark cycles at 28°C. All flasks were cultured under static conditions and were shaken by hand twice a day.

SMS was provided from Nippon Steel Company Oita Factory. The SMS composition was determined by x-ray fluorescence spectrometry (XRF) (ZSX101e, RIGAKU, Japan). Table 1 shows the chemical composition of the SMS, which exhibits the typical characteristics of arc furnace oxidizing slag.

2.2. Methods

2.2.1. Cell growth. Cell growth was determined by measuring the optical density at 556 nm using a spectrophotometer (U-1100 Spectrophotometer, HITACHI Ltd, Japan).

2.2.2. Metabolite analysis. The experiment examined the effect of 0, 50, 100, 500, 1,000, and 10,000 mg L^{-1} SMS during 60 days of cultivation.

The total lipid content was analysed using a Soxhlet apparatus following the method of Bligh and Dyer [13]. To quantify the intracellular carbohydrates, aliquots of 100 mg of lyophilized cells were added into a 2.5 M HCl solution (20 mL) and hydrolyzed in boiling water for 30 min. Then, the pH of the sample was adjusted to 7 using a NaOH solution (5 M) and water was added up to 50 mL. Finally, the mixtures were centrifuged for 5 min at 5000 × *g*, and the supernatant was used for carbohydrate determination using the DNS method [14].

2.2.3. SOD activity assay. Samples were harvested by centrifugation at 12,000 rpm for 5 min at 4°C and were then washed twice with sterile distilled water. Aqueous extracts of *S. platensis* M135 were prepared by homogenizing the microalgae (0.1–0.5 g fresh weight) and mixing with an equal volume of glass beads in 1.2 mL sterile distilled water at 4°C. The time for homogenization was optimized using a light microscope to verify complete cell breakage. The homogenates were then centrifuged at 12,000 rpm for 10 min at 4°C and then stored at −20°C until the superoxide dismutase (SOD) assay [15].

2.2.4. DNA profile of S. platensis M135. Variations in the DNA profile of S. *platensis* M135 were compared between the control and the different concentrations of SMS. Samples with different concentrations of SMS cultured for 30, 45, and 60 days were analysed, respectively. DNA of each sample was extracted using the ZR Plant/Seed DNA MiniPrep Kit for Random amplified polymorphic DNA based on polymerase chain reaction (RAPD-PCR). After the PCR reaction, the PCR product was analysed using agarose gel electrophoresis [16, 17].

3. Results and discussion

3.1. Growth promotion effect of SMS on S. platensis M135

Figure 1 shows the growth curves of *S. platensis* M135 grown with various concentrations of SMS. Growth promotion effects of SMS were obtained at all concentrations compared to the growth levels of the control without SMS. Figure 2 shows the growth promotion ratio of *S. platensis* with SMS based on the growth of control regarded as 100% at 60 days of culture. A growth promotion effect was observed

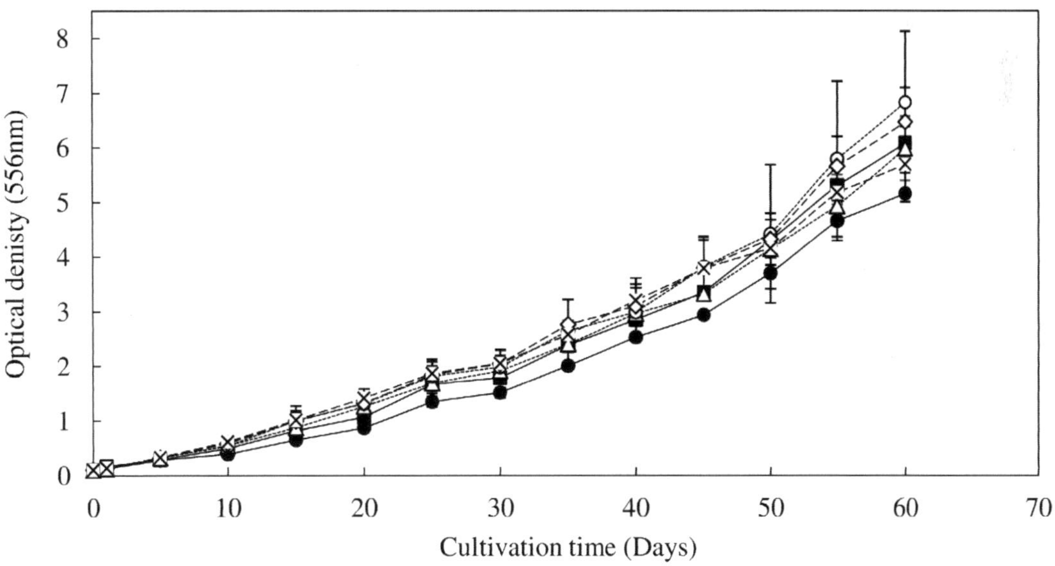

Figure 1. Growth curve of *S. platensis* M135 with steelmaking slag(SMS) (■: 50 mg L^{-1}, △: 100 mg L^{-1}, ○: 500 mg L^{-1}, ◇: 1,000 mg L^{-1}, and ×: 10,000 mg L^{-1}) and without SMS (●). The data are means ± standard deviations.

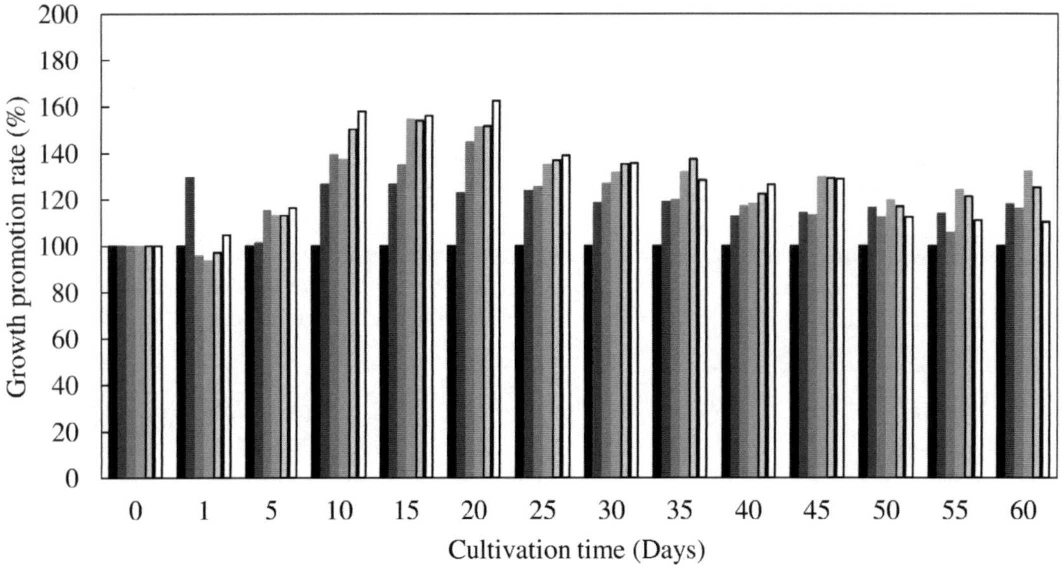

Figure 2. Growth promotion ratio based on the growth without steelmaking slag (SMS) as 100% (■) compared with SMS (■: 50 mg L^{-1}, ■: 100 mg L^{-1}, ■: 500 mg L^{-1}, ■: 1,000 mg L^{-1}, □: 10,000 mg L^{-1}).

after 5 days and the highest growth-promoting ratio of 1.62 was achieved with SMS of 10,000 mg L^{-1}. A growth promotion effect of 1.27 times greater than that with the control was obtained with 500 mg L^{-1} SMS at 60 days. The growth profile of this experiment agreed with the growth promotion effect of different iron forms with EDTA on *S. platensis* [18]. Thus, the iron dissolution from SMS is considered to be the growth promoter. Erosion of wustite and brownmillerite interfaces lead to Mg, Al, Si, and Ca dissolution from SMS slag to water [19]. Considering the composition of wustite [(Fe, Mg, Mn)O/MgO–FeO] and brownmillerite [$Ca_2(Al, Fe)_2O_5\cdot(4CaO\cdot Al_2O_3\cdot Fe_2O_3)$], which are contained in this SMS, Fe should also be eluted and available for *S. platensis* M135. Based on the solubility diagram for iron, the estimated form of iron eluted from SMS could be $Fe(OH)_4^-$. A time course of this dissolution might explain the growth promotion effect during culture. The growth promotion ratio was not proportional to the SMS concentration. Zhang *et al.* reported that the dissolved amount of Si and Fe decreases with increasing the ratio of slag/seawater, this is because iron dissolution is not linked to the slag concentration [20].

Table 1. Elemental composition of SMS.

Element	Fe	Ca	Mn	Si	Mg	Al	S	P
mass %	45.44 ±3.53	39.22 ±2.77	7.15 ±0.33	4.18 ±0.98	1.96 ±0.31	1.40 ±0.23	0.36 ±0.09	0.29 ±0.08

3.2. Metabolite analysis

Figure 3 shows the metabolite (carbohydrate and lipid) content of *S. platensis* M135 at 60 days of culture. SMS decreased the lipid content of *S. platensis* M135, but there was no change in the carbohydrate content from that of the control. The lipid and carbohydrate contents of the M135 strain are considered higher than those of other strains. The lipid content of *S. platensis* M135 of this study was comparable to a previously reported value [21], even though treated with SMS. While, another study with different microalgae *Chlorella*, the lipid content increased up to 7.27 times with 0.67 mg L^{-1} Fe^{3+} [10].

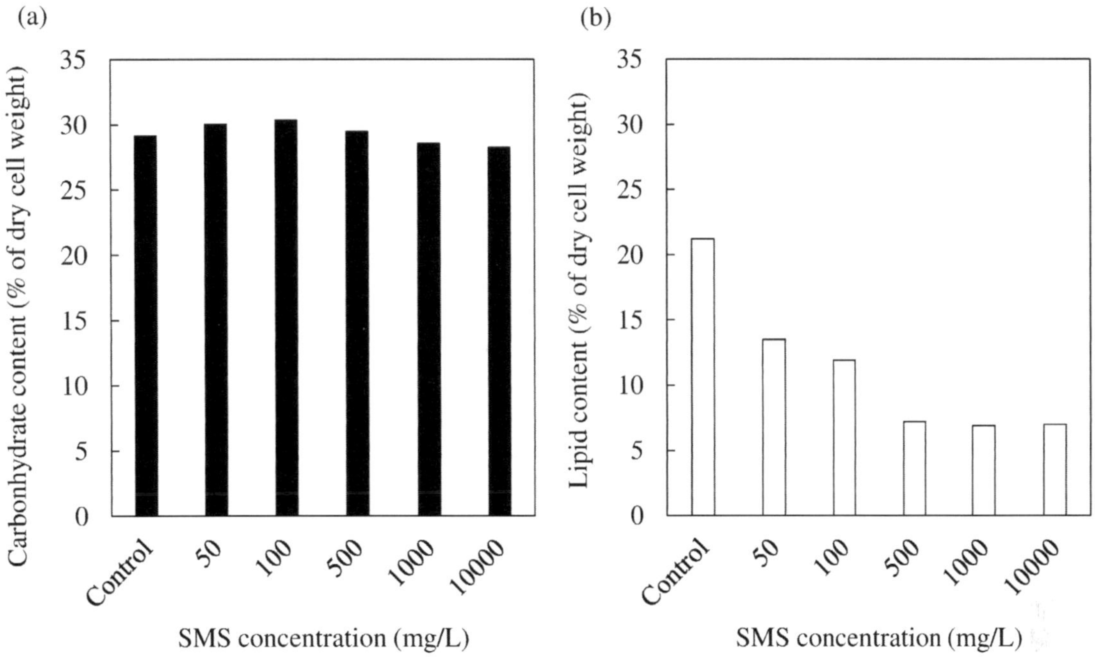

Figure 3. Carbohydrate (a) and lipid (b) content in the dry biomass of *S. platensis* M135 at 60 days of culture, respectively.

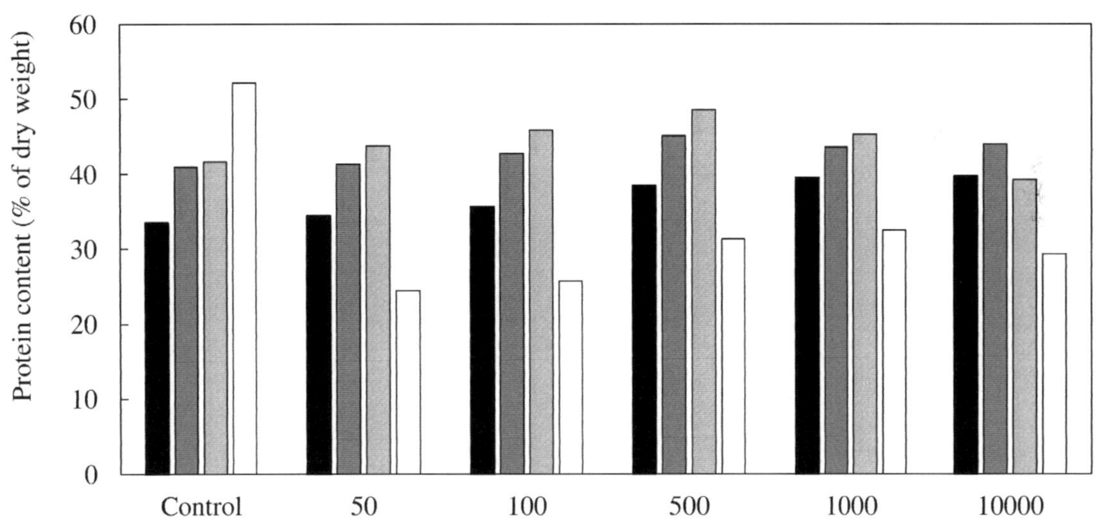

Figure 4. Protein content in the dry biomass of *S. platensis* M135 at 1 (■), 30 (■), 45 (■), and 60 days (□) of culture.

Figure 4 shows the protein content of *S. platensis* M135 at 1, 30, 45, and 60 days of culture. The protein content increased with concentration-dependent manner of SMS until 45 days, but then decreased from that with the control at 60 days. A slow change in the protein content occurred because of the iron form of SMS and its time consuming dissolution. Similarly, pigment-protein complex of phycocyanin content increased at the early stages of culture but decreased at the latter stages with different forms of iron and EDTA [18]. A chelating agent like EDTA is not provided when applying SMS. The strong chelator EDTA has been found to diminish the growth promotion effect of iron from SMS with compost [22]. Protein content increases should also be associated with iron dissolution. This idea is also supported by the increase in marine phytoplankton fluorescence intensity by iron from SMS (SMS diameter: 5–20 μm) [23]. Another explanation for the change in protein content is the pH increase. The highest amino acid content, corresponding to the protein content was obtained at pH 9 with *S. platensis*, but the amino acid content decreased with pH change [24]. The pH of the medium was affected by Ca and Mg from the SMS. Despite the continued buffering action of Mg, the Fe from SMS increased the protein content, while Ca release lead to a higher pH than that in the control, which also affected the protein content.

3.3. SOD activity

Figure 5 shows the SOD activity of *S. platensis* M135 at 1, 30, 45, and 60 days. SOD activity decreased with a time-dependent manner but showed an increasing trend in the control. The SOD activity was lower than that in the control at day 1, but it was higher at 30 days. Similarly, iron promoted the growth of *Heterosigma akashiwo* [25] and *Chattonella marina* [26] while reducing hydrogen peroxide production, which indirectly in accordance with our results of SOD activity.

What is the relationship between the SOD activity and microalgae growth with SMS? How can microalgae utilize inorganic iron? Possibly, the release of organic Fe chelators from various types of microalgae including *S. platensis* M135 help the uptake of iron from the medium. Based on a study of organic Fe chelator release from *Scenedesmus*, Benderliev postulated that the contact of inorganic Fe (II) and humic Fe (II) with the microalgae cell surface might result in the generation of strong oxidants such as superoxide radicals and hydroxyl radicals that tend to inhibit the release of chelators from cells [27]. The contact of Fe (III) with the cell's surface triggers another set of reactions that stimulate the release of chelators from cells. Thus, contact with iron from SMS might trigger organic Fe chelator release from *S. platensis* M135 and lead to growth promotion.

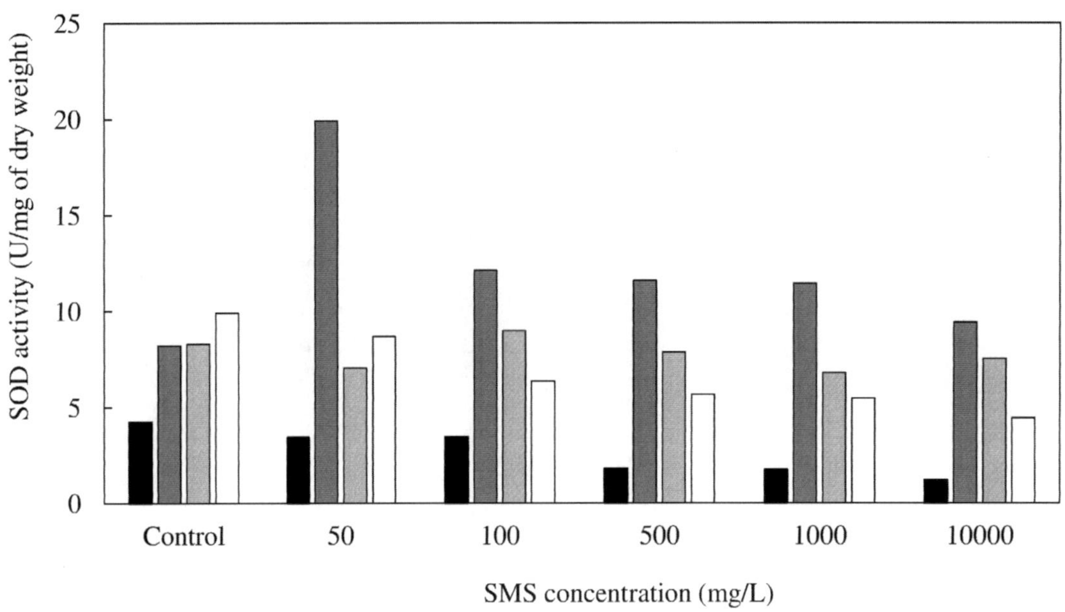

Figure 5. Superoxide dismutase (SOD) activity of *S. platensis* M135 dry biomass at 1 (■), 30 (■), 45 (■), and 60 days (□) of culture.

3.4. Effect of SMS on DNA profile of S. platensis M135

Figure 6 shows phylogenetic trees of six samples of *S. platensis* M135 cultured with 0, 50, 100, 500, 1,000, and 10,000 mg L^{-1} SMS for 30, 45, and 60 days. The degree of polymorphism in the samples treated with different concentrations of SMS increased with a time-dependent manner of 7%, 28%, and

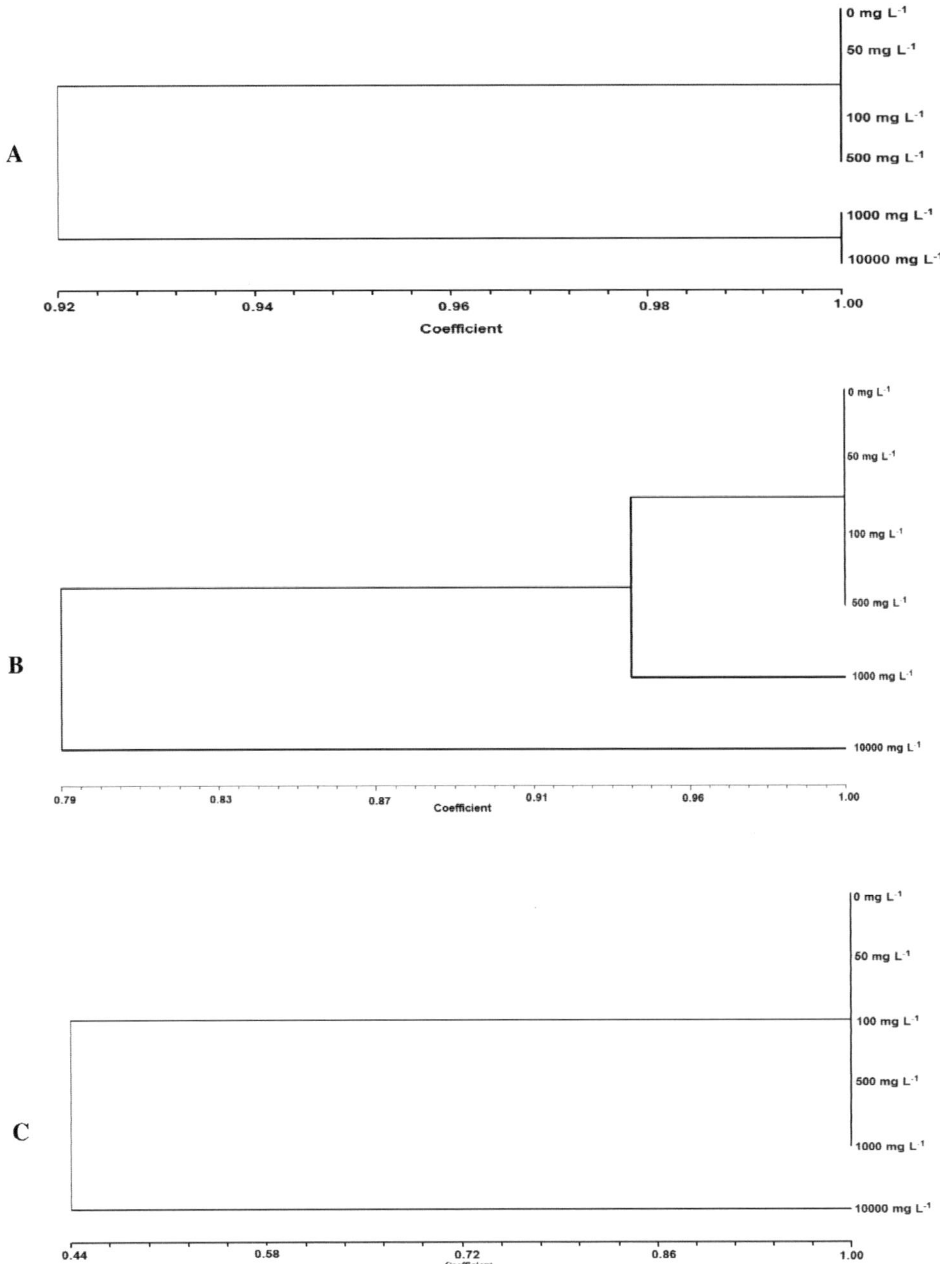

Figure 6. Phylogenetic tree of *S. platensis* M135 cultured with SMS concentration from 0, 50, 100, 500, 1,000 and 10,000 mg L^{-1} after 30 days of cultivation (A), 45 days of cultivation (B) and 60 days of cultivation (C).

33% at 30, 45, and 60 days, respectively. Genetic similarity coefficient of samples treated with 10,000 mg L^{-1} SMS decreased from 0.923 to 0.788(or 0.824) and then to 0.444, which differed from that of the samples treated with 0–1,000 mg L^{-1} SMS. For 0–1,000 mg L^{-1} SMS, after 60 days, the DNA levels are not different from those of the control. This may be an adaptive process of *S. platensis* M135 to lower concentrations of SMS.

4. Conclusions

SMS continuously promoted the growth of *S. platensis* M135 without largely affecting its metabolite and DNA profile. A growth promotion effect of 1.27 times greater than that with the control treatment was obtained with up to 500 mg L^{-1} SMS at 60 days of culture. The lipid content decreased in an SMS-concentration-dependent manner. The protein content increased in an SMS-concentration-dependent manner at 45 days of culture. No genetic damage was observed with up to 500 mg L^{-1} SMS, and recovery from genetic damage was observed with 1,000 mg L^{-1} SMS at 30 days of culture.

References

[1] Fujisawa T, Narikawa R, Okamoto S, Ehira S, Yoshimura H, Suzuki I, Masuda T, Mochimaru M, Takaichi S, Awai K, Sekine M, Horikawa H, Yashiro I, Omata S, Takarada H, Katano Y, Kosugi H, Tanikawa S, Ohmori K, Sato N, Ikeuchi M, Fujita N and Ohmori M 2010 *DNA Research* **17** 85
[2] Chen T, Zheng W, Wong Y S, Yang F and Bai Y 2006 *Bioresource Technology* **97** 2260
[3] Ak I 2012 *Aquaculture International* **20** 413
[4] Press release from Nippon Slag Association, 2003, http://www.slg.jp/e/images/Steel%20Slag%202014FY.pdf (Accessed 18 August 2015)
[5] Takahashi T and Yabuta K 2002 *NKK Technical Report-Japanese Edition-* **87** 43
[6] Miyata Y, Sato Y, Shimizu S and Oyamada K 2009 *JFE Technical Report* **13** 41
[7] Nakamura Y, Taniguchi A, Okada S and Tokuda M 1998 *ISIJ international* **38** 390
[8] Sugie K and Taniguchi A 2011 *ISIJ international* **51** 513
[9] Nagai T, Miki O and Okumura C 2014 *Journal of Water and Environment Technology* **12** 285
[10] Liu Z Y, Wang G C and Zhou B C 2008 *Bioresource Technology* **99** 4717
[11] Kadar E, Rooks P, Lakey C and White D A 2012 *Science of the Total Environment* **439** 8
[12] Sutak R, Botebol H, Blaiseau PL, Léger T, Bouget F Y, Camadro J M and Lesuisse E 2012 *Plant Physiology* **160** 2271
[13] Bligh E G and Dyer W J 1959 *Can. J. Biochem. Physiol.* **37** 911
[14] Miller G L 1959 *Analytical chemistry* **31** 426
[15] McCord J M and Fridovich I 1969 *Journal of Biological Chemistry* **244** 6049
[16] Hien H T M, Lan V T and Hong D D 2000 *Journal of Biology* **22** 269
[17] Hong D D, Hien H T M, Son P N, Bach N D and Dong N V 2002 *Journal of Science and Technology* **40** 161
[18] Sayin S and Doganay S 2011 *Fresenius Environmental Bulletin* **20** 1888
[19] Mombelli D, Mapelli C, Barella S, Gruttadauria A, Le S G and Garcia D E 2014 *Journal of Hazardous Materials* **279** 586
[20] Zhang X, Matsuura H and Tsukihashi F 2012 ISIJ International **52** 928
[21] Tokuşoglu Ö and Üunal M K 2003 *Journal of food science* **68** 1144
[22] Yamamoto M, Fukushima M and Liu D 2012 *ISIJ International* **52** 1909
[23] Haraguchi K, Suzuki K and Taniguchi A 2003 *ISIJ International* **43** 1461
[24] Ogbonda K H, Aminigo R E and Abu G O 2007 *Bioresource Technology* **98** 2207
[25] Twiner M J and Trick C G 2000 *Journal of Plankton Research* **22** 1961
[26] Liu W, Au D W T, Anderson D M, Lam P K S and Wu R S S 2007 *Journal of experimental marine biology and ecology* **346** 76
[27] Benderliev K 1999 *Bulg J Plant Physiol* **25** 65

Cascaded H-bridge multilevel inverter for renewable energy generation

Ravikant Pandey,Ravi Nath Tripathi,Tsuyoshi Hanamoto

Graduate School of Life Science and Systems Engineering, Kyushu Institute of Technology, 2-4 Hibikino, Wakamatsu, Kitakyushu 808-0196, Japan

Email:91ravikantpandey@gmail.com

Abstract. In this paper cascaded H-bridge multilevel inverter (CHBMLI) has been investigated for the application of renewable energy generation. Energy sources like solar, wind, hydro, biomass or combination of these can be manipulated to obtain alternative sources for renewable energy generation. These renewable energy sources have different electrical characteristics like DC or AC level so it is challenging to use generated power by connecting to grid or load directly. The renewable energy source require specific power electronics converter as an interface for conditioning generated power .The multilevel inverter can be utilized for renewable energy sources in two different modes, the power generation mode (stand-alone mode), and compensator mode (statcom).The performance of the multilevel inverter has been compared with two level inverter. In power generation mode CHBMLI supplies the active and reactive power required by the different loads. For operation in compensator mode the indirect current control based on synchronous reference frame theory (SRFT) ensures the grid operating in unity power factor and compensate harmonics and reactive power.

1. Introduction

Renewable energy sources like solar, hydro, wind, and biomass can be used for electrical power generation. Energy conversion of renewable energy sources with current technology in [1][2] that would offer dependable and efficient electricity facility .In recent years renewable energy system has become significant alternative of conventional electrical energy generation due to increase in fuel prices, limited reserves and environmental issues. The concern stood by conventional energy sources like non availability of fuels and climate change has been addressed in many countries and aiming to increase the development in the renewable energy system [3].Energy generated by the renewable sources can feed local loads and the surplus power can be transferred to the grid with the use of power electronics converter [4][5].The quality of electrical power transfer between renewable energy sources and grid depends on the quality of voltage/current[6].The voltage and current fed by the converter should be synchronized to grid with Harmonics level defined by IEEE and IEC standard. The integration of renewable source requires proper power control that plays important role for system efficiency and required output of the system. The available power from these renewable energy sources can be optimize and maximize by significant control of the system. Some renewable energy system may require power tracking control to track the maximum power point as in case of PV(Photovoltaic), wind power generating units. The system design requires the proper control, suitable converters, optimum cost considering all ecological factors.

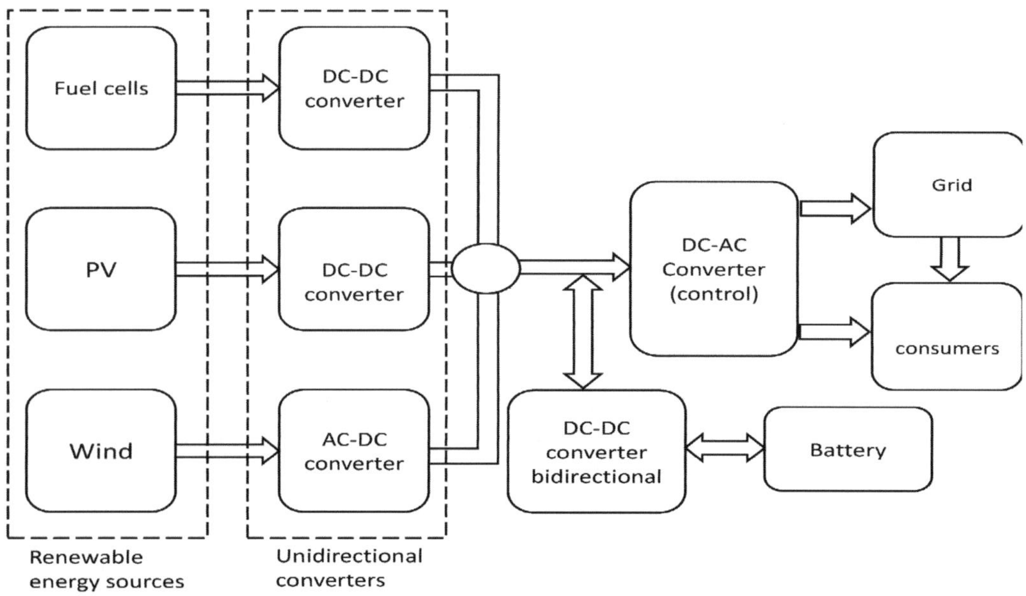

Fig.1 Topology of renewable energy generation system

The connection and significance of power electronics converters for renewable energy system consist of Fuel, PV and wind as distributed power generating unit is represented as block diagram in Fig.1.To obtain significant improvement in power quality the two level PWM(Pulse width modulation) converters are extended to multilevel inverters [7][8].Multilevel inverter for DC-AC conversion generates synthesized output voltage by chopping the DC voltage in to different levels [9].The multilevel inverter was designed in the earlier stage for generation of three level output voltage waveform ($+V_{dc}$,0,-V_{dc}). The inverter topology is extended to various new topologies. Three measure classification of multilevel inverter topologies are (i) Diode clamped MLI [10] (ii) capacitor clamped MLI [11](iii) cascaded H-bridge multilevel inverter (CHBMLI)[12]. The CHBMLI is more suitable for application of renewable energy system due to its modular structure and requirement of distributed DC sources. The efficiency of CHBMLI is better due to enhanced harmonics profile of stepped output wave with less component required in comparison to other MLIs. It is having other advantages like less dv/dt stress on load and less electromagnetic interference which make these type of power converters more appropriate [13].

In this paper integration of renewable energy system considered as DC source is investigated in which dc-ac conversion with respect to CHBMLI for stand-alone and grid connected mode is done. The CHBMLI is used with equal DC voltage source. The performance of multilevel inverter is compared and analyzed with respect to conventional two level inverter considering power quality of the system. The effectiveness of the multilevel topology is tested under power generation mode and in compensator mode for modelled and simulated in MATLAB/SIMULINK using power system tool box.

2. Cascaded H-bridge Multi level inverter

CHBMLI was introduced for motor drive application at initial stage. But in high voltage application due to the limited power rating of electronics switches, the conventional converters became very lossy and inefficient. So CHBMLI become significant for renewable energy system integration to electrical power grid for medium and high power application. The Cascaded H-bridge multilevel inverter provides lower cost, less electromagnetic interference, higher performance i.e. lower THD (Total harmonic distortion), and higher efficiency for grid voltage or current conditioning application in both shunt and series compensation. Each H-bridge of symmetrical CHBMLI requires a separate DC source of equal value so integration of different renewable energy sources is less complex in comparison with other multilevel inverter topology. The generated voltage through three phase symmetrical structure of CHBMLI is closer to sinusoidal waveform having reduced THD level in respect of conventional inverter. The quality of the output waveform depends upon the synthesized levelled output. So performance of CHBMLI can be improved by increasing number of H-bridge. The relation of number of H-bridges (n) and levelled output per phase (m) is m=2n+1. Due to limitation of switches and losses in the system, the number of levels can be increased according to the system requirement and application.

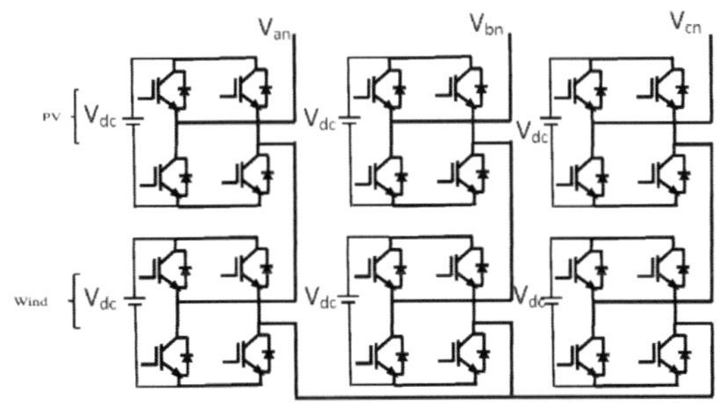

Fig.2 CHBMLI topology

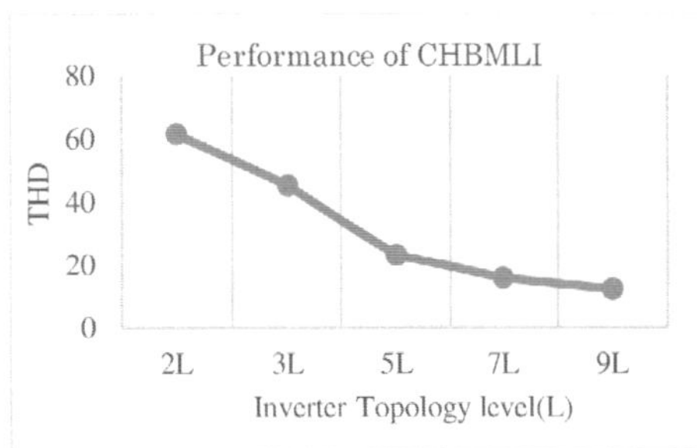

Fig.3 THD of MLIs

3. Performance of CHBMLI in Power generation mode

The performance of the CHBMLI for renewable energy sources is investigated for the linear load of 40 KVA 0.8 pf lagging connected to conventional two level and three level, five level, seven level cascaded h-bridge inverters.THD level for The output voltage waveform is illustrated as a graph in Fig.3 for above mentioned different level of inverter. The synthesized levelled output voltage and load current for corresponding 3,5 and 7 level inverter is shown in Fig.4,Fig5,Fig6.The load current for conventional two level of inverter is shown in Fig.7.On each bridge a separate DC source are used to supply 415 V (line to line) three phase load. It is observed that voltage levels are divided between 340 V (peak) and zero for 240 V rms voltage per phase. For three level inverter the voltage levels are -340V, 0V, 340V, for five level inverter the voltage levels are -340V,-170V, 0V, 170V, 340V, for seven level inverter the voltage levels are -340V,-170V,-85V,0V,85V,170V,340V.

It is observed that load is connected to different inverters are having same system parameter and performance is enhanced due to effect of multilevel topology. The output power of the inverter can be increased by integration of different small renewable energy sources which generate DC. This is one of the advantage of multilevel inverter which operates on distributed DC source system.

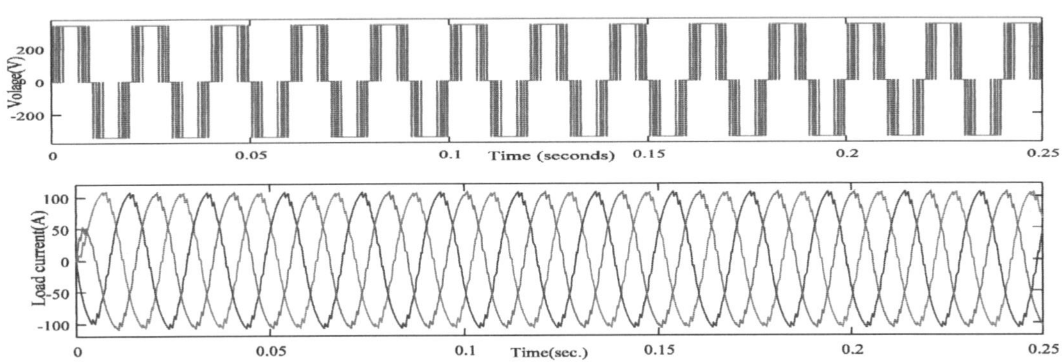

Fig.4 Three level inverter load voltage (phase a) and load current

Journal of Physics: Conference Series **704** (2016) 012020 doi:10.1088/1742-6596/704/1/012020

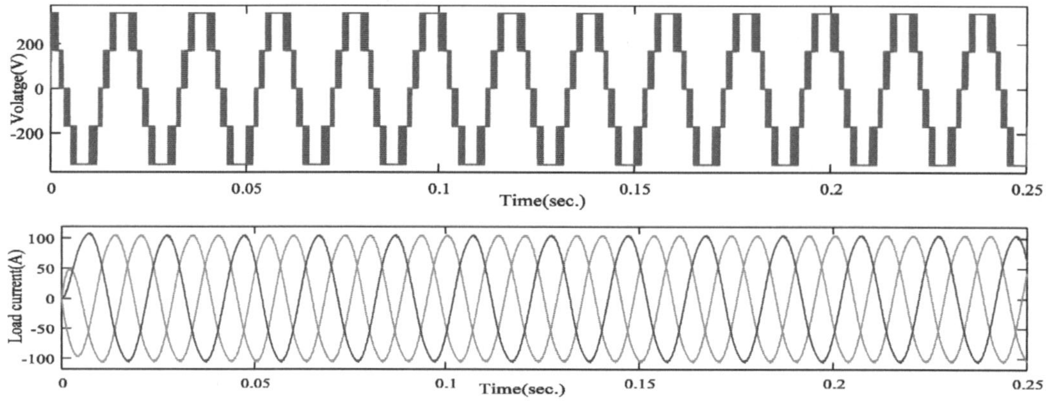

Fig.5 Five level inverter load voltage (phase a) and load current

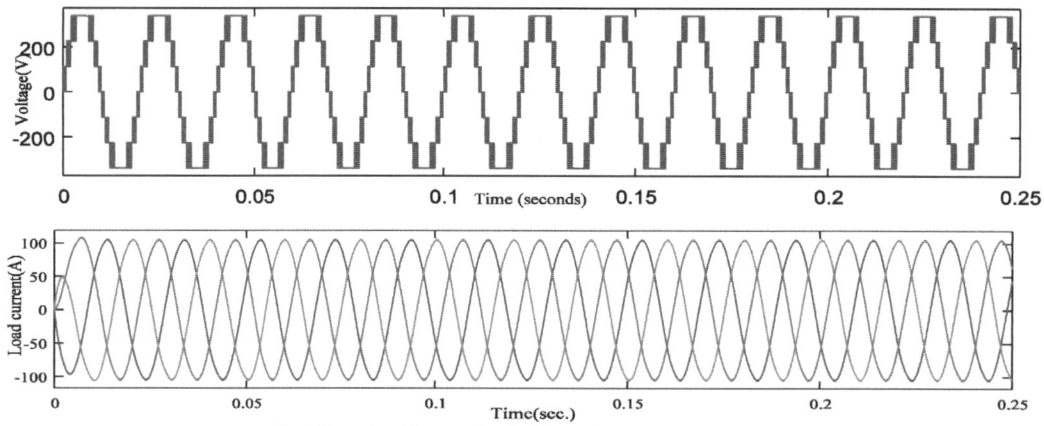

Fig.6 Seven level inverter load voltage (phase a) and load current

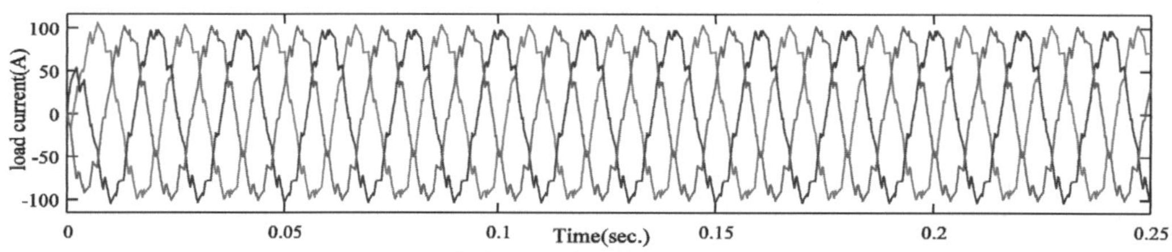

Fig.7 Conventional inverter load current

The output voltage and load current is showing improved waveforms with multilevel inverter and it can be observed in above illustrated waveforms. The THD of the output voltages and load current is significantly reduced with multilevel inverter even with less line inductance, small size of ripple filter interfacing the inverter and load as compared to conventional two level inverter. The THD of load current are 6.2%, 3.5%, 1.57%, 0.82%, for conventional two level, three level, five level, seven level inverters respectively under same load conditions.

4. Performance of CHBMLI in compensator mode

The integration of renewable energy sources using CHBMLI to grid can ensure the role of compensator as well as back-up of power. The power quality improvement is possible by compensating harmonics component injected by the load using suitable power control algorithm. The fundamental approach of control algorithm applied for renewable power system is to supply load reactive power and to compensate harmonics injected by load [14].The reference current generation is one of the measure aspect for compensator design and control and grid integrated renewable energy system with generalized control shown in Fig.8.

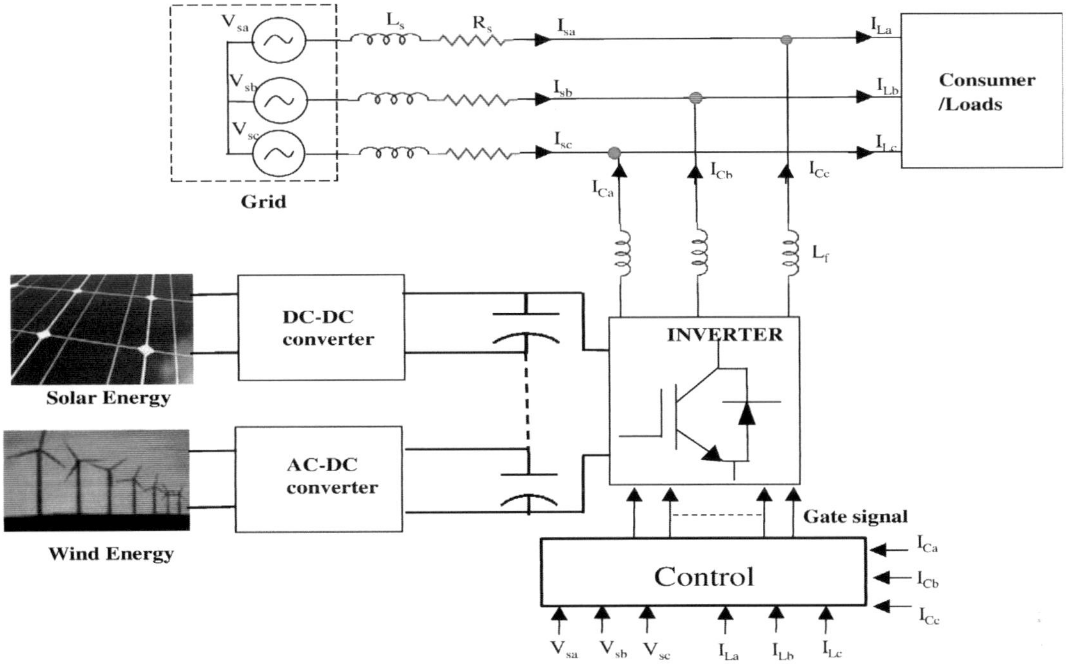

Fig.8 Grid integration of Renewable energy sources

At the consumer side, load is variant and could have different characteristics such as low power factor, Nonlinear. As a compensator the role of CHBMLI based renewable energy system is to reduce the reactive power supplied by the grid and harmonics injected by load up to its capacity to improve the operating power factor of the grid. Grid integrated renewable energy source analysed for linear load of 25 KVA at 0.8 pf and diode rectifier load of (10+7.45)Ω under compensation mode is assumed for performance evaluation of the system.The Control block in Fig.8 is responsible for synchronization of power generated from renewable system and grid using Phase locked loop(PLL) and multilevel inverter comprises of additional switching devices in comparison to conventional inverter.So sinusoidal pulse width modulation(SPWM) used for conventional inverter is extended for switching of multilevel inverter[15].In this paper PSPWM (Phase shift pulse width modulation) is used to control the multilevel inverter requires hierarchal switching scheme[16].For compensator reference current generation, algorithms like synchronous reference frame theory (SRFT), power balance theory, Instantaneous Power theory is introduced in [17].In this paper the SRFT based control is implemented for generation of reference current . The result shown in Fig.9 shows the impact of compensator to the grid current and

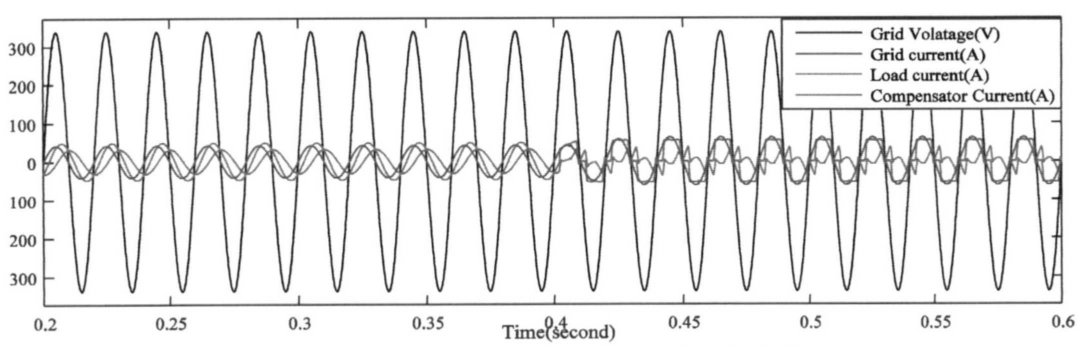

Fig.9 Grid voltage, current connected to linear load, non-linear load with compensator

voltage while load is not ideal (low p.f. and nonlinear) at the consumer end. At t=0 sec. the linear load was in operation with the system and at t=0.4 second it is switched to the non-linear load.The grid is operating under unity power factor linear for linear load condition as renewable system is able to supply reactive power of the load and also power factor is improved under nonlinear load condition.The load current THD for non-linear load is 24.6% and grid current THD is maintained at 3.58 %.

5. Conclusion

The CHBMLI topology is investigated and simulated for the renewable energy sources connected to three phase electrical power grid and power generation mode. The system is operating well and load current THD is maintained well within the standard limits for multilevel inverter and is significantly improved in comparison to two level inverter. In compensation mode, the grid power factor has been successfully improved for linear and non-linear load condition.The CHBMLI topology is working satisfactorily and suitable for integration of different renewable energy sources to grid.

References

[1] Trappey A.J.C., Trappey C.V.,Wang D.Y.C,.Li S.J and Ou J.J.R,*Proceedings of the 2014 IEEE 18th CSCWD* 716.
[2] Colak I., Kebalci E., 2013 *Electric Power Component and systems* **41** 31.
[3] Khare V., Nema S., Baredar P.,2016 *Renewable and sustainable Energy Reviews* **58** 13.
[4] Kakosimos P.,pavlou K., Kladas A.,S.Manias,2015 *Energy Conversion and Management* **89** 427.
[5] Mahalakshmi R., Thampatty K.C.S.,2015 *Procedia Technology-Smart Grid Technologies* **21** 636.
[6] Seetharaman A., Sandanaraj L.L.,2016 *Renewable and Sustainable Energy Reviews* **54** 1368.
[7] Lai J.S., Peng F.Z., 1996 *IEEE Transaction on Industry Application* **32** 509.
[8] Palanivel P. Dash S.S.,2011 *IET Power Electronics* **4** 951.
[9] Pandey R.,Chawla P.Tripathi R.N.,*Proceeding of 2015 IEEE RDCAPE* 27.
[10] Boumaaraf H.,Talha A.,Bouhali O.,2015 *Renewable and Sustainable Energy Reviews* **49** 1171.
[11] Lin B.R.,Huang C.H.,2004 *IEEE Proceeding of Electric Power* Applications **151** 555.
[12] Liang W., Donglai Z.,Yi W., Bin W., Aftab H.S.,2016 *IEEE Transaction on Power Electronics* **31** 3289.
[13] Latran M.B.,Teke A., 2014 *Renewable and sustainable Energy Reviews* **42** 361.
[14] Singh B.,Jayaprakash.Kothari D.P.,2011 *Electrical Power and Energy Systems* **33** 1109.
[15] Zeng Z., Yang H., Zhao R., Cheng C.,2013 *Renewable and Sustainable Energy Reviews* **24** 223.
[16] Rabinovici R., Baimel D, Tomasik J., Zuckerberger A., 2013 *IET Power Electronics* **6** 1516.
[17] Singh B., Solanki J., 2009 *IEEE Transaction on Industrial Electronics* **56** 2738.

Analysis of Dynamic Response of Permittivity in a Liquid Crystal Cell with Flow

R Ozaki[1], K Matsuura[1], K Kadowaki[1], T Q Duong[2] and H Moritake[2]

[1]Department of Electrical and Electronic Engineering and Computer Science, Graduate School of Science and Engineering, Ehime University, 3 Bunkyo-Cho, Matsuyama, Ehime 790-8577, Japan
[2]Department of Electrical and Electronic Engineering, National Defense Academy, 1-10-20 Hashirimizu, Yokosuka, Kanagawa 239-8686, Japan

E-mail: ozaki.ryotaro.mx@ehime-u.ac.jp

Abstract. Microwave and millimeter-wave devices including a liquid crystal (LC) are increasingly attractive for the use in adaptive and controllable devices. Various types of microwave phase shifters having an LC transmission line have been studied (e.g. microstripline, coplanar waveguide, and rectangular waveguide). In conventional microwave devices, the response time after removal of voltage is slow because the LC layer in the devices is usually thick. In this study, the time response of the LC permittivity is studied experimentally and theoretically. Experimentally measured response curves having two time constants for decay are examined by using Frank's continuum theory with and without the backflow effect.

1. Introduction

With the increasing use of mobile communications, satellite broadcasting, and intelligent transport systems, liquid crystal (LC) materials have become increasingly attractive for use in adaptive and controllable devices in the microwave and millimeter-wave fields. LCs have various anisotropies, such as permittivity, refractive index, and viscosity, owing to their molecular shape [1]. The electrically control of the LC permittivity is a basic concept of the active devices for microwave and millimeter-wave. Various types of microwave phase shifters having an LC transmission line have been studied, for example, microstripline, coplanar waveguide, and rectangular waveguide [2-5]. We have also proposed the LC microstripline and coplanar waveguide phase shifters [6,7]. In the former studies, the effective dielectric birefringence of an LC layer, the variation of the phase shift, and the response time of microwave phase shifters have been examined experimentally and theoretically.

LC configuration and its dynamics are key roles in the study of the time response. In particular, the dynamics of LC molecular motions is very important because a slow response time of microwave devices is a major barrier for practical use. The typical cell thickness of the LC microwave phase shifters is several ten microns that is a cause of the slow response. Relaxation process governs the molecular reorientation after removal of voltage. In general, the response of the relaxation process can be fitted with an exponential curve. However, we observed a response curve having two time constants for decay in experiments of a microwave phase shifter. In this paper, we have experimentally and theoretically studied the dynamic response of permittivity in typical LC cells to investigate the response curve with two time constants.

2. Experiments of time response of LC permittivity

LC cells with different thicknesses were used to examine time responses of permittivity caused by molecular reorientation. The LC cells with thicknesses of 12.7, 19.0, and 44.8 μm were prepared with polyethylene terephthalate films. The thicknesses were estimated by measuring the capacitances of the cells with an impedance analyzer (4294A, Agilent) before infiltrating LC into the cells. In the cells, indium-tin oxide (ITO) glass substrates were spin-coated with polyimide (AL-1254, JSR) and the polymide layer was rubbed unidirectionally to obtain a planar cell. The LC cells were filled with a nematic LC 4-cyano-4A-pentylbiphenyl (5CB) (K-15, Merck).

A rectangular voltage at a frequency of 1 kHz supplied from a function generator (AFG3022, Tektronix) was applied to the cells via a power amplifier (4010, NF). The amplitudes of the voltages for all cells were set to be an electric field of 2.0 V/μm. That is, the amplitudes of the voltages for the 12.7-, 19.0-, and 44.8-μm cells were 2.54, 3.80, and 8.96 V, respectively. The permittivity of the LC cell at 100 kHz was measured with the impedance analyzer and a lock-in amplifier (SR844, Standford research systems).

Figure 1 shows the measured time-response waveforms of the LC permittivity after removing the applied voltage. Before 0 second, the electric field of 2.0 V/μm had been applied to each cell, and then the electric field was immediately decreased to zero at 0 second. In figure 1(a), the permittivity of the 12.7-μm cell shows an exponential decay curve with one time constant. On the other hand, in the 19.0- and 44.8-μm cells, the response curves are slightly different from the decay curve of the 12.7-μm cell. In the 19.0- and 44.8-μm cells, the permittivity quickly decreases at the moment that the voltage is zero, and then they exponentially decrease with a large time constant. In other word, the decay curve of the 12.7-μm cell has one time constant, while the decay curves of the 19.0- and 44.8-μm cells have two time constants.

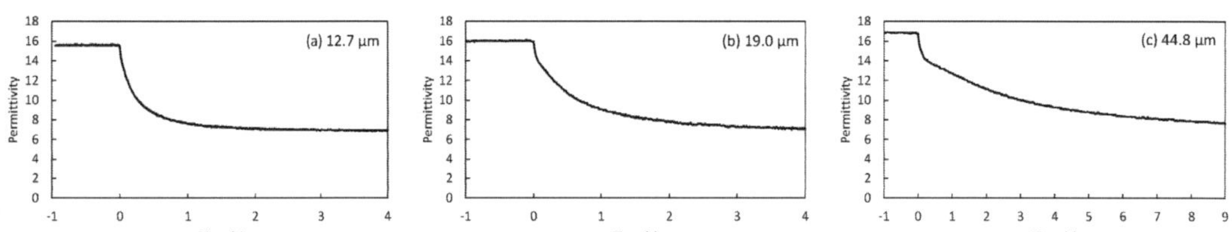

Figure 1. Measured response waveforms of permittivity for the 12.7-, 19.0-, and 44.8-μm LC cells.

3. Calculations of time response of LC permittivity

3.1 Continuum theory

LC director configurations in the cells were calculated using Frank's continuum theory [1,8] to investigate the characteristics of the permittivity response. Using one-constant approximation, the free energy density *f* for nematic LC is given by

$$f = \frac{1}{2}K\left(\frac{d\theta}{dz}\right)^2 - \frac{1}{2}\varepsilon_0 \Delta\varepsilon E^2 \sin^2\theta, \tag{1}$$

where K is a Frank elastic constant, θ is the director angle, ε_0 is the vacuum permittivity, $\Delta\varepsilon$ is the anisotropy of LC permittivity, and E is the electric field. The director configuration can be obtained by substituting f into the Euler-Lagrange equation. The torque balance equation for LC is written as

$$\gamma\frac{\partial\theta}{\partial t} = K\frac{\partial^2\theta}{\partial z^2} + \varepsilon_0 \Delta\varepsilon E^2 \sin\theta\cos\theta, \tag{2}$$

where γ is the rotational viscosity of LC.

Figure 2 shows the director configurations in the LC cells. In these calculations, we assumed that the pretilt angle of the LC is 1.0° and that the boundary condition is weak anchoring. We set to be 0 second when the applied voltage was removed after the application of voltage with a sufficient time. Almost all of the directors direct to 90° at 0 second. After removing the applied voltage, the director angles gradually decrease with time. The spatial distributions of the LC permittivity were calculated using the calculated director angles. The time responses of permittivity were obtained by averaging the spatially distributed LC permittivity. Figure 3 shows the response waveforms of the LC permittivity calculated using Frank's continuum theory. We obtained three decay curves which show a good agreement with the experimental data in terms of time scale. However, the decay curves for the 19.0- and 44.8-μm cells do not have two time constants.

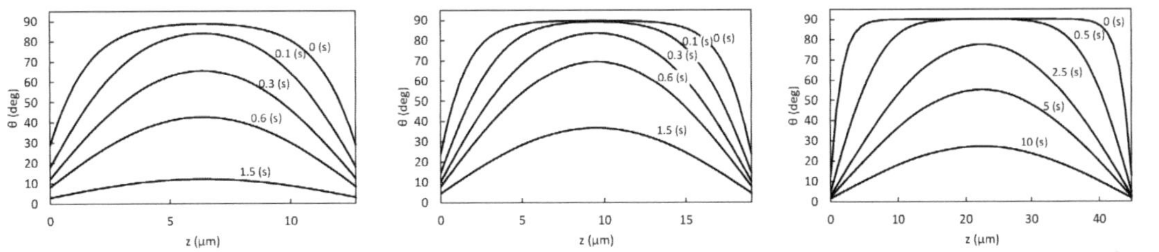

(a) 12.7 μm (b) 19.0 μm (c) 44.8 μm

Figure 2. LC director configurations calculated using Frank's continuum theory.

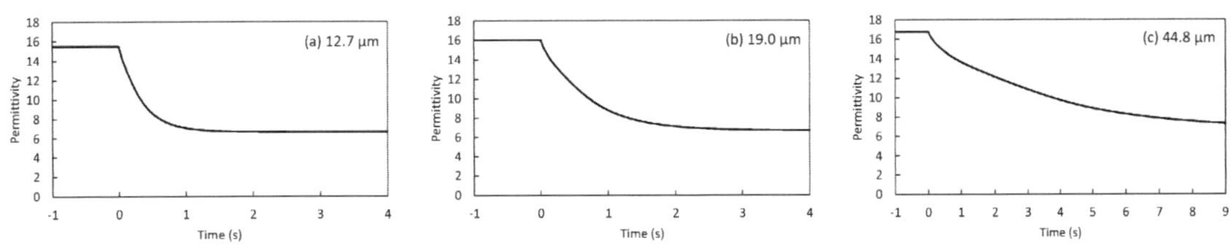

Figure 3. Calculated response waveforms of LC permittivity using Frank's continuum theory.

3.2 Dynamics with backflow

In section 3.1, we calculated the LC director configurations and the time-response curves of the LC permittivity using Frank's continuum theory. The calculated decay curves had the same time scale in comparison with the experimental results. However, the calculated curves did not have two time constants which were the characteristic features of the experimental results. To investigate more details, we calculated the LC director dynamics taking into account flow of LC molecules. Backflow of LCs is a well-known phenomenon in display applications [9]. The torque balance equation including the backflow effect is written as

$$\gamma\frac{\partial\theta}{\partial t} = K\frac{\partial^2\theta}{\partial z^2} + \varepsilon_0\Delta\varepsilon E^2\sin\theta\cos\theta - (\alpha_3\cos^2\theta - \alpha_2\sin^2\theta)\frac{\partial v_x}{\partial z}, \quad (3)$$

where v_x is a flow velocity, α_2 *and* α_3 are the Leslie coefficients. In addition to equation (3), the Navier-Stokes equations must be solved to obtain v_x. In this study, equation (3) was numerically solved using the Crank-Nicolson method.

Figure 4 shows the flow velocities in the cells obtained by solving the Navier-Stokes equations. Flow occurs in the LC cells at the moment that the voltage is removed. Therefore, the flow velocities at 0.01 seconds are the largest in each graph. Since the boundary effect from the glass substrate is relatively weak in a thick cell, the flow velocity in the 44.8-μm cell is larger than those in the 12.7- and 19.0-μm cells. The flow is symmetry in the LC cell, and the flow velocity at the center of the cell is always zero.

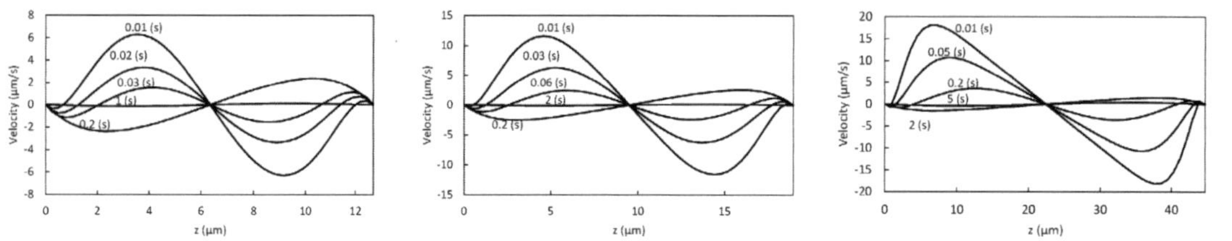

(a) 12.7 μm (b) 19.0 μm (c) 44.8 μm

Figure 4. Calculated flow velocities in the LC cells.

The LC director configurations calculated using Frank's continuum theory with backflow are shown in figure 5. After removing the applied voltage, the director angle decreases to 0°. Note that the director angle in the center area temporarily increased from 90°. This is caused by the backflow effect and this is the difference with the results shown in figure 2. The 44.8-μm cell shows the largest increase in director angle because the flow velocity is also the largest. The spatial distributions of the LC permittivity were calculated as the same manner in section 3.1, and the time responses of the permittivity were obtained by averaging the spatially distributed LC permittivity. Figure 6 shows the response waveforms of the LC permittivity calculated using Frank's continuum theory including the backflow effect. It is noteworthy that the time-response curves in figures 6(b) and 6(c) have two time constants. In contrast, the time-response curve in figure 6(a) appears to be an exponential curve with one time constant. The reason can be explained as follows: The LC permittivity at 90° is the permittivity along the long molecular axis that is the largest permittivity for the LC. When the backflow occurs in the cell, the director angle increases beyond 90°, and then the angle decreases to 0° via 90°. Therefore, in the molecular reorientation, the LC permittivity at the center area decreases, increases, and decreases again. This is the reason that the time-responses in the 19.0-μm and 44.8-μm

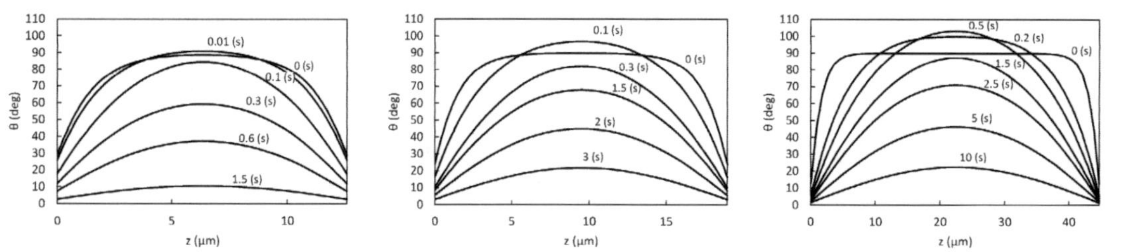

(a) 12.7 μm (b) 19.0 μm (c) 44.8 μm

Figure 5. LC director configurations calculated using Frank's continuum theory with backflow.

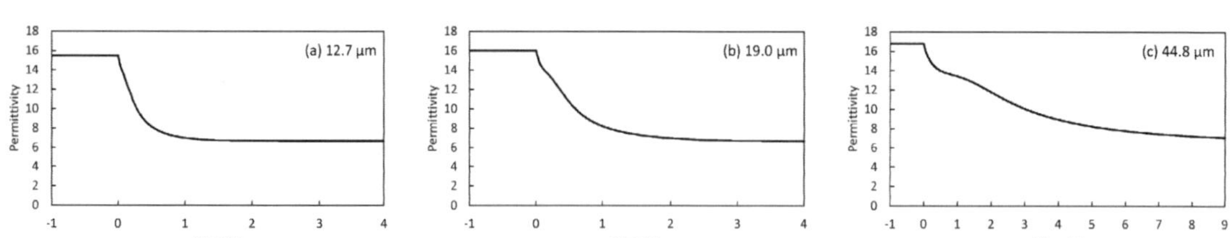

Figure 6. Calculated response waveforms of LC permittivity using Frank's continuum theory with backflow.

cells do not show an exponential decay curve with one time constant.

4. Conclusions

The time response of LC permittivity was experimentally and theoretically investigated. The measured responses of the 19.0- and 44.8-μm cells after removal of voltage showed an exponentially decay curve with two time constants. The responses calculated by Frank's continuum theory did not show a decay curve with two time constants but showed an exponential decay curve with one time constant. We also calculated the LC director dynamics with taking into account of the backflow effect. The response curve including the backflow consisted of an exponential decay with two time constants. The calculations showed a qualitative agreement with the experimental results. The two time constants of a decay curve were explained by the backflow effect.

Acknowledgments

This work was supported by JSPS KAKENHI Grant Number 26420290.

References

[1] de Gennes P G and Prost J 1993 *The Physics of Liquid Crystals, 2nd ed.* (Oxford Science, Oxford)
[2] Lim K C, Margerum J D and Lackner A M 1993 *Appl. Phys. Lett.* **62** 1065
[3] Utsumi Y, Kamei T, Saito K and Moritake H 2005 *IEEE Transactions on Microwave Theory and Techniques* **53** 3345
[4] Moritake H, Morita S, Ozaki R, Kamei T and Utsumi Y 2007 *Jpn. J. Appl. Phys.* **46** L519
[5] Nose T, Yanagihara S, Sato Y, Ito R and Honma M 2008 *Jpn. J. Appl. Phys.* **47** 8483
[6] Utsumi Y, Bach N T, Kamei T, Ozaki R and Moritake H 2009 *IEICE Trans. Electron.* **J92-C** 778 (In Japanese)
[7] Nguyen T, Nguyen B T, Ozaki R and Moritake H 2013 *Jpn. J. Appl. Phys.* **52** 121701
[8] Frank F C 1958 *Discuss. Faraday Soc.* **25** 19
[9] Berremen D W 1975 *J. Appl. Phys.* **46** 3746

Supramolecular Nanocomposites: Dispersion of Zero-, One- and Two-dimensional Nanoparticles in Discotic Liquid Crystals

Sandeep Kumar

Raman Research Institute, C.V. Raman Avenue, Bangalore – 560 080, India

E-mail: skumar@rri.res.in

Abstract. Discotic liquid crystals are emerging as novel nanomaterials useful in many device applications. Recently their hybridization with various zero-, one- and two-dimensional metallic and semiconducting nanoparticles has been realized to alter and improve their thermal, supramolecular and electronic properties. In this article, we have overviewed the work carried out in our laboratories on the dispersion of various metallic, semiconducting and carbon nanoparticles in discotic liquid crystals. First a brief introduction of self-organizing supramolecular liquid crystalline materials is presented with an emphasis on discotic liquid crystals. This is followed by the description of various discotic liquid crystal-nanoparticle hybrid systems. A number of discotic liquid crystals, functionalized nanoparticles and their nanocomposites were prepared and studied by spectroscopic and analytical tools. The dispersion of such functionalized nanomaterials in columnar matrix enhances the physical properties such as, conductivity, photoconductivity, absorbance, etc., significantly without disturbing the supramolecular properties.

1. Introduction

Supramolecular systems are formed from molecular systems with the help of many types of interactions, such as π–π interactions; dipolar and quadrupolar interactions; van der Waals interactions, charge transfer; hydrogen bonding and metal coordination. Supramolecular assemblies are extremely important in biological science; no life is possible without the self-assembly of lipid molecules into bilayers assembly within the cell membrane. Numerous Biosystems such as, DNA, collagen, enzymes, etc., are formed by the self-organization of small or large molecules using molecular recognition and supramolecular interactions [1]. Supramolecular assemblies formed due to non-covalent interactions in neat systems as well as in solvents are also very important in materials science. Clathrate compounds, host-guest complexes, molecular sieves, hydrogen-bonded complexes of guanidine, cyanuric acid, etc., are a few examples.

Liquid crystals (LCs) belong to one of such systems where supramolecular interactions play a crucial role in the formation of LCs and in the determination of their mesomorphic properties. Liquid crystalline state is a well-defined thermodynamically stable state of matter possessing the properties of both crystalline state and isotropic liquid state. Thus they combine both the order and mobility which is the basic principle for self-organization and structure formation in living systems. As the physical properties of LCs fall in between those of the crystalline solid and amorphous liquid states, they are referred to as intermediate phases or mesophases. Some mesophases are very sensitive to small external forces, such as electric and magnetic fields, which is the basis for their application in information display. Many bio-molecules, such as carbohydrates, lipids, nucleic acids, have been found to exist in various liquid crystalline phases [2]. The formation of LC phases by nano DNAs (small fragment of DNA) indicates the significant role of LCs in the evolution of life in the prebiotic world [3].

LCs are ubiquitous in our present days life in the form of liquid crystal display devices (LCDs). Currently more numbers of LCDs exist than the people on earth. The LCDs industry is a multi-billion dollar industry. There are several other applications/potential applications of LCs such as, temperature-sensing (medical thermometers), solvent in chemical reactions, in chromatography, in spectroscopy, in holography, etc. Depending on how the mesophases are obtained, liquid-crystalline phases can be classified in two types, thermotropic and lyotropic. Lyotropic liquid-crystalline phases are obtained by mixing amphiphilic compounds in suitable solvents, quite often water, over a range of concentrations and temperatures. These are extremely important in biosciences. The thermotropic liquid-crystalline phases are obtained by heating a pure compound or a mixture of compounds or by cooling an isotropic liquid. LCs have also been classified as rod-like (calamitic) and disc-like (discotic), depending on the shape of the mesogenic molecules. This article deals only with the discotic liquid crystals.

The phenomena of liquid crystallinity in organic molecules was discovered in 1888 [4] and till 1977 all the mesogenic molecules reported were calamitic (rod-shaped). However, in 1977 a group of Indian scientists at the Raman Research Institute, Bangalore found that compounds with disc-like molecular shape are able to form liquid crystals [5]. They prepared some hexaesters of benzene (Figure. 1) and from optical, thermal and X-ray studies, it was confirmed that these materials form a new class of LCs in which molecules are stacked one on top of the other in columns and these molecular columns self-assemble in a hexagonal arrangement (Figure. 1).

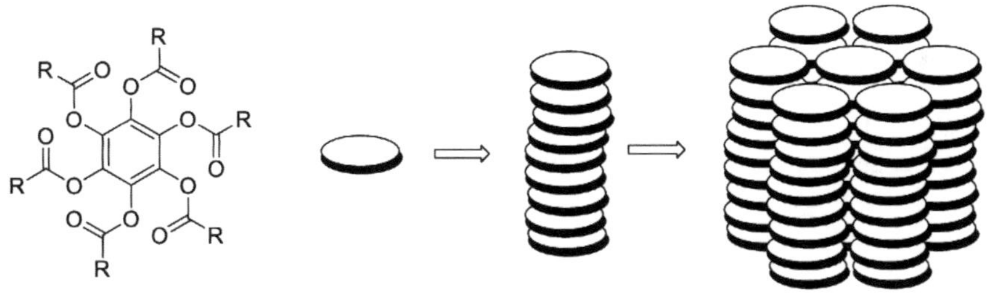

Figure 1. Molecular structure of hexaalkanoyloxybenzene, the first DLC and its self-organization in the hexagonal columnar mesophase.

Discotic mesogens form primarily two types of mesophases; (i) nematic having only orientational order of molecules and (ii) columnar, having both orientational and positional order [6]. Both types of mesophases are very important for various device applications. Nematic DLCs can be polymerized to form negative birefringence films which have been commercialized as compensation foils to enlarge the viewing angle of commonly used LCDs [7,8]. They can also be used as active component in LCDs [8-10]. A majority of DLCs are derived from polycyclic aromatic cores having strong π-π interactions and, therefore, arrange themselves in columns. The separation in between two molecules (core–core separation) in a columnar mesophase is usually of the order of 3.5 Å. Therefore, there is considerable overlap of p-orbitals. As flexible long aliphatic chains surround the core, the separation between two columns (intercolumnar separation) is usually 20–40 Å, depending on the lateral chain length. Therefore, interactions between neighbouring molecules within the same column are much stronger than interactions between neighbouring columns. Consequently, the migration of charge or energy in these materials is expected to be quasi one-dimensional. Accordingly, conductivity along the columns has been reported to be several orders of magnitude greater than in the perpendicular direction [11]. Therefore, the columnar mesophase may be described as molecular wires. Due to this unique property, DLCs have been extensively studied for the energy and charge migration and their device applications such as, one-dimensional conductors, photoconductors, light emitting diodes, photovoltaic solar cells, field-effect transistors and gas sensors have been sought [12].

2. Nanomaterials

The past two decades have experienced an explosive development in the fields of nanostructured materials such as, metallic nanoparticles (NPs) and nanorods (NRs), semiconducting quantum dots (QDs), carbon nanotubes (CNTs), fullerenes and graphene. Materials that have at least one dimension in the range of 1–100 nm are defined as nanomaterials. These materials possess unique properties, which differ from those of bulk material and from individual atoms. For instance, when the size of metals decreases to one billionth of a meter, their conductivity decreases significantly and they become semiconductors; some metal NPs, but not bulk metal, possess antibacterial activity; the optical absorption (color) and fluorescence properties of metal NPs are very different than in the bulk materials [13, 14]. The nanoscience and nanotechnology have received great interest in nearly every field of science. The potential applications of nanomaterials in the fields of energy, computing, optics, catalysis, biosciences and medical sciences have been extensively discussed [15].

3. Liquid crystal nanoscience

Liquid crystals are organic nanoparticles usually in the range of 2-6 nm. Like nanomaterials, LCs also possess unique properties that differ from those of either crystals or liquids. Due to the emergence of this strong link between nanoscience and LCs, a new term, 'Liquid Crystal Nanoscience,' has evolved [16, 17]. Assembling metallic, semiconducting or carbon NPs into hierarchical self-assembling structures is a fascinating way to prepare advanced functional materials. The synergetic relationship of LCs possessing order and mobility at the molecular and supramolecular levels with the excellent electronic properties of various zero-dimensional (0-D), one-dimensional (1-D) and two-dimensional (2-D) metallic, semiconducting and carbon NPs may produce advanced functional materials for next generation electronic and opto-electronic applications [18-20]. Efforts have been made to alter the physical properties of LCs and NPs either by attaching liquid crystalline molecules covalently to NPs or by adding a low concentration of NPs to a liquid crystalline matrix. In the following sections, we will present an overview of our work on the dispersion of various functionalized NPs in DLCs.

4. Zero-dimensional nanoparticles in discotic liquid crystals

A three dimensional object when reduces to nanometer regime, is classified as zero-dimensional (0-D) object. Thus, NPs having all dimensions below100 nm are referred to as 0-D NPs. We have recently looked the dispersion of spherical 0-D GNPs (GNPs), CdSe and CdTe QDs in DLCs.

4.1. Spherical gold nanoparticles

The existence of GNPs is known since ancient times. However, the scientific evaluation of colloidal gold started by the classical work of Michael Faraday in 1857 [21]. With the availability of commercial electron microscope to well characterize nano-sized objects in conjunction with the synthesis of stable monolayer protected GNPs [22], the field grown rapidly. The Brust–Schiffrin simple synthesis of monolayer protected GNPs in organic solvent reported in 1994 [22] is perhaps the most popular method for producing smaller gold particles. In this method, a strong reducing agent, e.g., sodium borohydride was used to reduce gold(III) salt in the presence of an alkanethiol capping agent to produce 1–3 nm GNPs. We followed this method to prepare simple hexanethiol-protected GNPs as well as discotic functionalized GNPs [23, 24]. The synthesis is shown in Scheme 1.

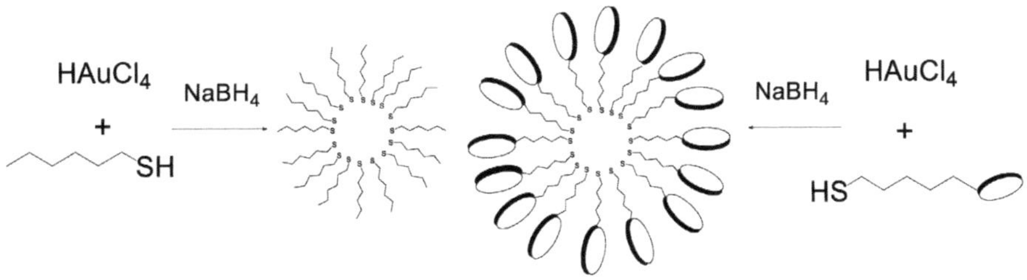

Scheme 1. Synthesis of alkane- and discotic-functionalized GNPs

Hexanethiolate-capped gold clusters of 1.6 nm core diameter with an average composition of $Au_{140}[S(CH_2)_5CH_3]_{53}$ were prepared by reducing $HAuCl_4$ with $NaBH_4$ in presence of hexanethiol. These NPs were fully characterized and dispersed in three different DLCs namely hexaxexylthio-triphenylene (HHTT), hexapentyloxytriphenylene (HPT) and hexakis(4-nonylphenylethynyl)benzene (H9B) [23]. HHTT is known to exhibit a highly ordered helical phase in addition to a hexagonal columnar mesophase [25]; HPT exhibits an ordered hexagonal columnar phase [26] and H9B displays a discotic nematic phase [27]. A few binary mixtures of hexanethiolat-functionalized GNPs with various DLCs were prepared by mixing the two components in a volatile solvent (dichloromethane) followed by removal of solvent and drying the sample under vacuum. Though, GNPs could not be well dispersed (a phase segregation was observed) in discotic nematic sample, the columnar phases of HHTT and HPT could well disperse the NPs in their matrices. These nanocomposites were analyzed by differential scanning calorimetry (DSC), polarizing optical microscopy (POM), X-ray diffraction (XRD), and conductivity studies. From experimental observations, it is attributed that GNPs are dispersed in the columnar phase evenly with a possibility of the formation of self-assembled superlattices of GNPs in the helical phase of HHTT [28]. An increase in the conductivity of the system was observed due to the presence of GNPs in the matrix [23, 29]. We also observed a decrease in orientational order parameter (S) and an increase in relaxation time (τ) for disc motion with GNPs in columnar phase [30]. We have also looked the dispersion of hexanethiolate-functionalized GNPs in a lyotropic liquid crystalline phase and measured the dimensions of the hexagonally ordered phases both with and without nanoparticles using SAXS. The results show that the GNPs are dispersed in between the columns, essentially retaining the column dimensions [31].

These initial experiments clearly indicated that small spherical small GNPs can be easily dispersed in the columnar matrix of DLCs in low concentration without disturbing their mesomorphism. To look the effects of peripheral substitution on the GNPs, we designed GNPs fully covered with triphenylene discotic mesogens. Such a functionalization may yield liquid-crystalline GNPs or may have better compatibility with discotic host [24]. Thus, reduction of $HAuCl_4$ in the presence of thiol-functionalised triphenylene derivative yields discotic decorated GNPs (Scheme 1). These triphenylene-functionalised GNPs self-assemble into hexagonal patterns on the surface. This could be due to the strong π–π interactions between the triphenylene ligands of adjacent NPs. Shen *et al.* have also reported similar self-assembly of triphenylene covered GNPs on surfaces [32]. These discotic-functionalised GNPs could be easily dispersed in large quantities into the columnar phase of a related triphenylene DLC namely hexaheptyloxy-triphenylene (H7TP), owing to their chemical compatibility with one another. Remarkably, DC conductivity measurements show an enhancement of electrical conductivity by six orders of magnitude (Figure 2) under ambient conditions upon doping a very small amount (1%) of discotic-functionalized GNPs. We believe that GNPs are randomly dispersed in the columnar mesophase (Figure 2) and the enhancement in the conductivity could be due to the formation of charge transfer complexes between electron deficient NPs and electron rich DLC.

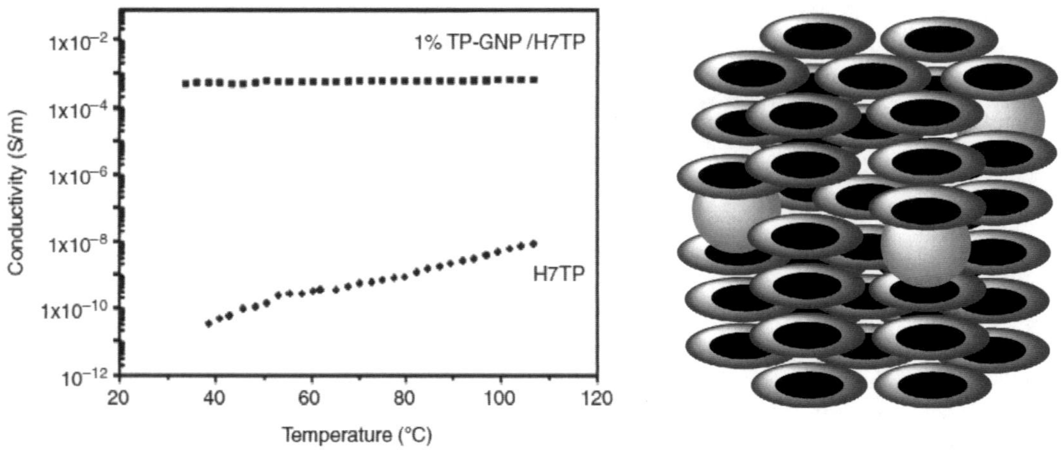

Figure 2. DC conductivity values as a function of temperature for 1% TPGNPs mixed with H7TP and neat H7TP and representation of the dispersion of NPs in a columnar metrics.

Later, similar observations were made by Holt *et al.*[33]. They also observed six orders of magnitude enhancement in electrical conductivity in the hexagonal columnar phase of a triphenylene-based DLC upon doping with methylbenzene thiol-covered GNPs [33]. The enhancement in conductivity due to formation of small chain-like aggregates of GNPs was proposed.These observations are highly significant from device application point of view. Further studies on structure–property relationship are desirable to fully understand the mechanism of charge migration in these nanocomposites and to further explore their potential applications.

4.2. Quantum dots in discotic liquid crystals

Quantum dots (QDs) are zero-dimensional semiconductor inorganic NPs.They have been extensively studied in recent times due to their potential applications in the life sciences and materials science [34-37]. We looked the dispersion of CdSe and CdTe QDs in DLCs [38,39]. These QDs were prepared as reported [40, 41] with some minor modifications and characterized by transmission electron microscopy (TEM), UV-Vis spectroscopy and photoluminescence spectroscopy. The TEM results confirm the synthesis of spherical particles with uniform sizes and shapes. Nanocomposites of both CdSe and CdTe QDs with triphenylene DLCs were prepared by mixing two components in chloroform under sonication followed by vacuum drying. These nanocomposites were studied by polarized optical microscopy, differential scanning calorimetry, X-ray and DC conductivity. Results were quite similar to GNPs dispersion in DLCs. The dispersion of both CdSe and CdTe quantum dots in DLCs show higher conductivity compared to the pure compound. The doping of quantum dots into discotic liquid crystals does not disturb the nature of mesophase with different concentration variations but minor shift in the transition temperatures is observed. The enhanced conductivity makes QDs doped DLCs as potential candidates for various device applications such as, thin film transistors, LED's, and organic solar cell.

Over all, it may be concluded that spherical metallic or semiconducting zero-dimensional NPs can be evenly distributed in the columnar matrix. The dispersion of NPs enhances the conductivity of the system upto six orders of magnitude under ambient conditions, making these nanocomposites suitable for application in many devices.

Journal of Physics: Conference Series **704** (2016) 012022 doi:10.1088/1742-6596/704/1/012022

5. One-dimensional nanoprticles in discotic liquid crystals

Nanoparticles in which one dimension is out of the nanoscale while the others are in the nanoscale (1-100 nm), are termed as 1-D nanomaterials. These include nanotubes, nanorods, nanowires, etc. We have studied the dispersion of carbon nanotubes (CNTs) and gold nanorods (GNRs) in DLCs.

5.1. Carbon nanotubes in discotic liquid crystals

CNTs were discovered by Iijima in 1991 [42]. CNTs are all-carbon hollow cylinders of graphite with a high aspect ratio and are primarily two types; single-wall CNTs (SWNTs) which are formed by rolling a single sheet of graphene and, multi wall CNTs (MWNTs) formed by a few sheets of graphene. CNTs have been extensively studied due to their exciting electrical, mechanical and thermal properties [43-47]. Biomedical applications of CNTs have also been sought [48-50]. The functionalization of SWNTs with discotic molecules may generate liquid crystalline SWNTs, or will at least have better dispersibility in DLCs, with this idea, we functionalized SWNTs with triphenylene discotic mesogens [51]. Thus, acid-purified carboxylic acid-bearing SWNTs and hydroxyl-functionalized triphenylene derivatives were prepared and coupled via classical esterification reaction to yield discotic-functionalized SWNTs (f-SWNTs; Scheme 2). The product was well characterized by spectral, thermal and microscopic analyses. Nanocomposites of f-SWNTs and a DLC namely hexabutyloxy-triphenylene (H4TP) were prepared by mixing two components in dichloromethane under sonication followed by removal of solvent under vacuum. It is noteworthy that a very large amount (10%) of CNTs can be easily dispersed in DLC. This could be because of high compatibility of two systems. DLC-f-SWNTs nanocomposites were investigated using POM, differential scanning calorimetry and XRD. Results indicate insertion of the SWNTs in the supramolecular ordering of the columnar phase occupying the space between the disc columns (Scheme 2).

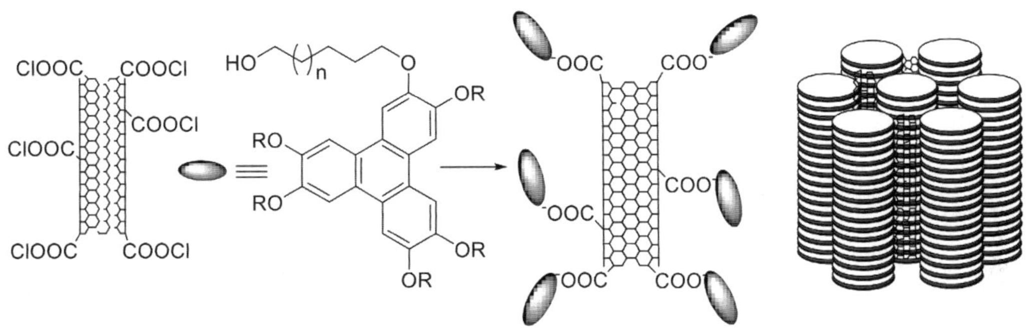

Scheme 2. Synthesis of discotic-functionalized SWNTs and their dispersion in columnar matrix

We have also studied the dispersion of non-discotic functionalized but processable simple octadecylamine-functionalized SWNTs in the columnar phases of triphenylene and rufigallol-based discotic monomers and polymers [52]. POM, DSC and XRD results indicate that these can also be dispersed in various discotic liquid crystalline monomers and polymers albeit in low concentration. Molecular compatibility of discotic-functionalized SWNTs due to π–π interactions of triphenylene molecules surrounding the SWNTs with the columnar phase forming triphenylene molecules likely stabilize the dispersion of f-SWNTs. Lee at al. dispersed pristine SWNTs in discotic ionic LCs derived from triphenylene core [53]. These nanocomposites exhibit anisotropic conducting properties.

5.2. Gold nanorods in columnar mesophase

The tunability of the surface plasmon band offers several applications of GNRs in bioscience and materials science [54, 55]. Unlike spherical GNPs, 1-D GNRs display two surface plasmon bands: a weak transverse band, similar to spherical GNPs, and a strong longitudinal band in the near-infrared region. The position of the longitudinal band red-shifts as the aspect ratio increases. GNRs with different aspect ratios can be prepared via a number of methods. However, the seed-growth methods developed by Murphy [56] and El-Sayed [57] have received much attention because of their simplicity and high yield of well-defined monodispersed GNRs. Accordingly we prepared GNRs with an aspect ratio of 2.7 (about 15 nm wide and 40 nm long). These GNRs were passivated with dodecanethiol. Dispersion of GNRs in a triphenylene DLC namely, hexapentyloxy-triphenylene (HPT) was made by mixing dichloromethane solution of two components followed by removal of solvent. Thus, a number of composites were prepared and characterized by UV–Vis spectroscopy, POM, DSC, XRD and conductivity studies. Results indicate the alignment of GNRs along the columnar axis. The GNRs doped sample exhibits much higher conductivity than the undoped sample.

5.3. Gold nanorods embedded in discotic nanoribbons
When a solution of DLCs in a nonpolar solvent is treated with a polar solvent, discotic nanoribbons composed of discotic molecules with the π planes stacking in the direction of the long axis of the ribbon are formed. Xiao *et al.* reported that a hexabenzocoronene-based DLC organises into molecular stacks and that these stacks self-organise into cables or fibres [58]. Such nanoribbons have huge application potential in electronic devices. A field effect transistor was constructed using an isolated nanofiber. Discotic triphenylene, perylene and decacyclene molecules have also been used to develop nanowires which can be used for explosive sensing [59-61]. Metallic nanorods embedded discotic nanowires may provide better 1D conducting properties, with this idea, we looked into the formation of nanoribbons of undoped as well as GNRs doped triphenylene discotics. Thus a chloroform solution of HPT or HPT-GNR composites when treated with methanol, produces discotic nanoribbons. The dark field scanning transmission electron microscopy (STEM) image of 5% GNR-doped HPT nanoribbons (Figure 3) clearly demonstrate that in the nanoribbons formed by HPT-GNR composite, the GNRs are aligned with its long axis parallel to length of the ribbon.

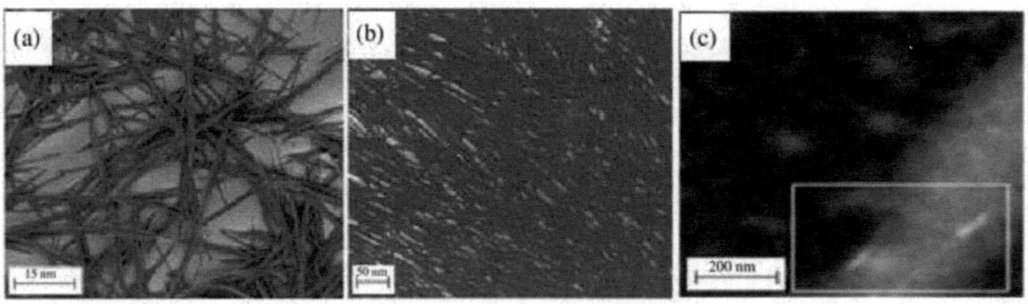

Figure 3. SEM images of randomly oriented fibres of HPT (a); fibres formed by 1% GNR-HPT (b) and dark field STEM images of gold nanorods embedded in HPT nanoribbons (c). The bright spots in the fibres are due to gold nanorods embedded in the ribbon with their longitudinal axis parallel to length of ribbon [62].

Discotic molecules primarily self-assemble into columns; these columns then self-organize into hexagonal structures which grow into ribbon like structures. Similar treatment of DLC-GNR composites results in the formation of nanorods embedded discotic ribbons. These composites exhibit enhanced conductivity owing to insertion of metallic nanorods in ribbon like structures and, therefore, could be interesting materials for devices like field effect transistors.

6. Two-dimensional nanoparticles in columnar mesophase
A large number of DLCs are derived from polycyclic aromatic hydrocarbons (PAH) such as, benzene, triphenylene, coronene, etc. Graphene is the largest PAH. With its two-dimensional structure and fascinating physical properties, graphene has emerged as a potential candidate for various device applications [63-67]. In order to understand the effects of graphene dispersion in DLCs, we prepared

processable functionalized reduced graphene oxide. Graphene can be prepared via physical as well as chemical routes. The chemical route is more suitable for the preparation of a substantial amount of the materials. Oxidation of graphite following the Hummers method [68] produces graphene oxide (GO) with epoxy, hydroxyl and carboxyl groups. Carboxyl groups present at the edges are converted to more reactive acid chlorides which can be coupled with long chain amines. Thus, octadecylamine-functionalized graphene (f-graphene) was prepared as reported by Haddon et al. [69]. During the treatment of GO with thionyl chloride, to generate acid chloride groups, reduction of GO occurs, with some defects, and f-graphene is formed [70]. Formation of the desired f-graphene was confirmed by IR spectroscopy, elemental analysis, and X-ray diffraction. To disperse the functionalized graphene in DLC, a room temperature DLC namely 1,5-dihydroxy-2,3,6,7-tetrakis(3,7-dimethyloctyloxy)-9,10-anthraquinone (RTAQ) was prepared [71]. Dispersion of f-graphene in DLC was carried out by sonicating a dichloromethane solution of DLC and f-graphene for 30 min followed by removal of the solvent under vacuum. These DLC–graphene nanocomposites were analysed by UV-Vis spectroscopy, POM, DSC, XRD, Raman spectroscopy, elemental analysis, conductivity studies, SEM and cryo-SEM analysis [72]. Results indicate that graphene can be easily dispersed in low concentration in DLC without disturbing the parent mesophase. The doping of f-graphene stabilizes the discotic mesophase. This is due to enhanced local ordering of discotics on the surface of f-graphene owing to the π–π stacking between their aromatic cores. The Cryo-SEM images of composite having 1% of f-graphene in DLC (Figure 4) clearly show a layered structure unlike that of pure DLC. Layers observed in Cryo-SEM images of the composite are more ordered than those of f-graphene. Discotic molecules are arranged in the interlayer spacing of the f-graphene sheets. The system is stabilized due to π–π interaction between aromatic cores of graphene and discotic molecules. Upon closely observing the Cryo-SEM images of the composites, one can see domains of molecular columns present on f-graphene layers (Figure. 4). Thus the system forms a "pillar-roof" type molecular building where discotic columns act as pillars and graphene as roof on these columns. The f-graphene-DLC composites show higher conductivity compared to pure DLC. The f-graphene layers act as bridges across defects resulting in more ordered mesophase and thus efficient charge transport. The creation of such mutually ordered sandwich like self-assembly of DLCs may widen the applicability of these nanocomposites in materials science.

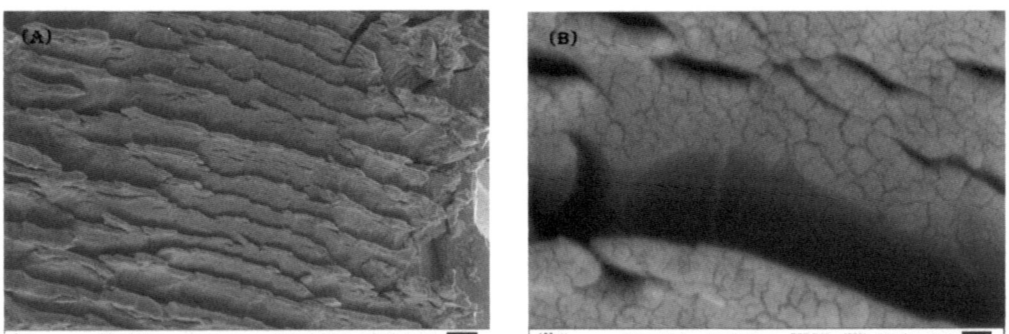

Figure 4. (A) Cryo-SEM image of the composite also showing a layered structure and enlarged portion of (A) showing graphene layers covered by a layer of discotics (B) [72].

We have also looked the dispersion of graphene functionalized on the surface [73]. Thus, thiolated graphene oxide was prepared by reacting GO with various thiols. Thiophenol and hexadecanethiol are used separately for functionalization of GO. Nanocomposites were prepared by mixing two components in dichloromethane followed by vacuum drying. Naocomposites were well characterized using POM, DSC, XRD, elemental analysis and SEM analysis. The results indicate that the dispersion of thiolated graphene in DLC causes destabilization of columnar mesophase. This could be because of the functionalization of graphene on the surface which disturbed the efficient packing of discotic molecule in a columnar phase. However, the thiol-functionalized graphene samples self-assemble onto gold surface due to strong gold-thiol interactions. This monolayer film of graphene acting as blanket on the gold surface has potential application in various devices applications such as sensors, organic electronic devices etc.

7. Summary and outlook

The field of liquid crystal nanoscience is derived from two fascinating research fields; liquid crystals and nanoscience. Discotic nanoscience is a part of LC nanoscience involving only DLCs and NPs. It is emerged in the scientific literature very recently but has made significant advances during the past decade. A variety of DLC-NP hybrid systems have been realized and studied because of their interesting properties. These hybrid systems may find applications in a wide range of devices, including photovoltaic solar cells, thin film transistors, photoconductors and so on. DLCs have been covalently attached to various NPs. These discotic functionalized NPs as well as other functionalized NPs have been dispersed in DLCs. The zero-dimensional NPs are randomly dispersed in the columnar phase; one-dimensional NPs go along the columnar direction and the two-dimensional on the surface of columns. The dispersion of these NPs does not affect the mesomorphic properties significantly but improve physical properties significantly. For example, the electrical conductivity of the nanocomposites increases by several orders of magnitude compared with the pure discotic system. These nanocomposites may provide the next generation of advanced functional materials for many device applications.

8. References

[1] Kumar S 2011 Chemistry of discotic liquid crystals: from monomers to polymers (Boca Raton, FL: CRC Press).
[2] Hamley I W 2010 *Soft Matter* **6** 1863.
[3] Nakata M, Zanchetta G, Chapman B D, Jones C D, Cross J O, Pindak R, Bellini T and Clark N A 2007 *Science* **318** 1276.
[4] Reinitzer F. 1888 Beiträge zur Kenntniss des Cholesterins. Monatsch Chem. **9** 421 (English Translation),1989, Contributions to the knowledge of cholesterol, *Liq Cryst.* **5** 7.
[5] Chandrasekhar S, Sadashiva B K and Suresh K A 1977 *Pramana* **7** 471.
[6] Kumar S 2006 *Chem. Soc. Rev.* **35** 83.
[7] Kawata K. 2002 *Chem Rec.* **2** 59.
[8] Bisoyi H K and Kumar S. 2010 *Chem Soc Rev.* **39** 264.
[9] Chandrasekhar S, Prasad S K, Nair G G, Rao D S S, Kumar S and Manickam M 1999 Euro Display '99, 19th international Display Research Conference, late-news 9.
[10] S. Kumar S and S. K. Varshney S K 2000 *Angew. Chem., Int. Ed.* **39** 3140.
[11] Balagurusamy V S K, Prasad S K, Chandrasekhar S, Kumar S, Manickam M and Yelamaggad C V. 1999 *Pramana.* **53** 3.
[12] Sergeyev S, Pisula W and Geerts Y 2007 *Chem. Soc. Rev.* **36** 1902.
[13] Daniel M C and Astruc D. 2004 *Chem. Rev.* **104** 293.
[14] Burda C, Chen X, Narayanan R and El-Sayed M A. 2005 *Chem. Rev.* **105** 1025.
[15] Rao C N R, Muller A and Cheetham A K. 2004 (eds) The Chemistry of Nanomaterials: Synthesis, Properties and Applications (Wiley-VCH, Weinheim,).
[16] Bisoyi H K and Kumar S. 2011 *Chem. Soc. Rev.* **40** 306.
[17] Garbovskiy Y A and Glushchenko A 2011 *Solid State Phys.* **62** 1.
[18] Stamatoiu O, Mirzaei J, Feng X and Hegmann T 2012 *Top. Curr. Chem.* **318** 331.
[19] Nealon G L, Greget R, Dominguez C, Nagy Z T, Guillon D, Gallani J L and Donnio B 2012 *Beilstein J. Org. Chem.* **8** 349.
[20] Lagerwall J P F and Scalia G. 2012 *Current Appl. Phys.* **12** 1387.

[21] Faraday M. 1857 *Phil Trans R Soc Lond.* **147** 145.
[22] Brust M, Walker M, Bethell D, Schiffrin D J and Whyman R. 1994 *Chem Commun.***7** 801.
[23] Kumar S and Lakshminarayanan V 2004 *Chem. Commun.* 1600.
[24] Kumar S, Pal S K, Kumar P S and Lakshminarayanan V 2007 *Soft Matter* **3** 896.
[25] Marguet S, Markovitsi D, Millie P, Sigal H and Kumar S. 1998 *J Phys Chem B.* **102** 4697.
[26] Kumar S and Manickam M 1997 *J Chem Soc Chem Commun.* 1615.
[27] Kohne B amd Praefcke K 1987 *Chimia.* **41** 196.
[28] Vijayaraghavan D and Kumar S 2009 *Mol. Cryst. Liq. Cryst.* **508** 101.
[29] Mishra M, Kumar S and Dhar R 2014 ***RSC Adv.*** **4** 62404.
[30] Supreet, Pratibha R, Kumar S and Raina K K 2014 *Liq. Cryst.* **41** 933.
[31] Kumar P S, Pal S K,Kumar S and Lakshminarayanan V 2007 *Langmuir* **23** 3445.
[32] Shen Z, Yamada M and Miyake M 2007 *J Am Chem Soc.* **129** 14271.
[33] Holt L A, Bushby R J, Evans S D, Burgess A and Seeley G 2008 *J Appl Phys.* **103** 63712.
[34] Liu L, Miao Q and Liang G 2013 *Materials* **6** 483.
[35] Shirasaki Y, Supran G J, Bawendi M G and Bulovic V 2013 *Nat. Photon* 7 13.
[36] Kamat P V 2012 *Acc. Chem. Res.* **45** 1906.
[37] Semonin O E, Luther J M and Beard M C 2012 *Mater. Today.* **15** 508.
[38] Kumar S and Sagar L. 2011 *Chem. Commun.* **47** 12182.
[39] Kumar M and Kumar S 2015 *RSC Adv.* **5** 1262.
[40] Zlateva G, Zhelev Z, Bakalova R and Kanno I 2007 *Inorg Chem.* **46** 6212.
[41] Wuister S F, Swart I, Driel F V, Hickey S G and Donega C D M 2003 *Nano Lett.* **3** 503.
[42] Iijima S 1991 *Nature.* **354** 56.
[43] De Volder M F L, Tawfick S H, Baughman R H and Hart A J 2013 *Science* **339** 535.
[44] Park S, Vosguerichian M and Bao Z 2013 *Nanoscale.* **5** 1727.
[45] Scarselli M, Castrucci P and De Crescenzi M 2012 *J Phys Condens Matter.* **24** 313202.
[46] Dai L, Chang D W, Baek J and Lu W 2012 *Small* **8** 1130.
[47] Endo M 2012 *Jpn J Appl Phys.* **51** 040001.
[48] Saifuddin N, Raziah A Z and Junizah A R 2013 *J Chem.* **2013** 676815.
[49] Peretz S and Regev O 2012 *Curr Opin Colloid Interface Sci.* **17** 360.
[50] Kamalha E, Shi X, Mwasiagi J I and Zeng Y 2012 *Macromol Res.* **20** 891.
[51] Kumar S, Bisoyi H K 2007 *Angew Chem Int Ed.* **46** 1501.
[52] Bisoyi H K and Kumar S 2008 *J. Mater. Chem.* **18** 3032.
[53] Lee J J, Yamaguchi A, Alam M A, Yamamoto Y, Fukushima T, Kato K, Takata M, Fujita N and Aida T 2012 *Angew Chem Int Ed.* **51** 8490.
[54] Murphy C J, Thompson L B, Chernak D J, Yang J A, Sivapalan S T, Boulos S P, Huang J, Alkilany A M and Sisco, P N 2011 *Curr. Opin. Colloid Interface Sci.* **16** 128.
[55] Huang X, Neretina S and El-Sayed M A 2009 *Adv. Mater.* **21** 4880.
[56] Jana N R, Gearheart L and Murphy C J 2001 *Adv. Mater.* **13** 1389.
[57] Nikoobakht B and El-Sayed M A 2003 *Chem. Mater.* **15** 1957.
[58] Xiao S, Tang J, Beetz T, Guo X, Tremblay N, Siegrist T, Zhu Y, Steigerwald M and Nuckolls C 2006 *J Am Chem Soc.* **128** 10700.
[59] Wang H, Xu X, Kojtari A and Ji H 2011 *J Phys Chem C.* **115** 20091.
[60] Che Y, Yang X, Liu G, Yu C, Ji H, Zuo J, Zhao J and Zang L 2010 *J. Am. Chem. Soc.* **132** 5743.
[61] Wang H, Xu X, Li L, Yang C and Ji H 2011 *Micro Nano Lett.* **6** 763.
[62] Avinash BS, Lakshminarayanan V, Kumar S and Vij JK 2013 *Chem Commun.* **49** 978.
[63] Li D, Muller MB, Gilje S, Kaner RB and Wallace C G 2008 *Nat. Nanotechnol.* **3(2)** 101.
[64] Dikin D A, Stankovich S, Zimney E J, Piner R D, Dommett G H B and Evmenenko G, 2007 *Nature* **448(7152)** 457.
[65] Geim A K and Novoselov K S 2007 *Nat. Mater.* **6(3)** 183.
[66] Zhang Y B , Tan Y W , Stormer H L and Kim P 2005 *Nature* **38 (7065)** 201.

[67] Novoselov K S , Geim A K, Morozov S V, Jiang D, Zhang Y and Dubonos S V 2004 *Science* **306(5696)** 666.
[68] Hummers S W and Offeman E R 1958 *J. Am. Chem. Soc.* **80** 1339.
[69] Niyogi S, Bekyarova E, Itkis M, McWilliams J, Hamon M and Haddon R 2006 *J. Am. Chem. Soc.* **128** 7720.
[70] Kumar N, Choi H J, Shin Y, Chang D, Dai L and Baek J B 2012 *ACS Nano* **6** 1715.
[71] Bisoyi H K and Kumar S 2008 *J. Mater. Chem.* **18** 3032.
[72] Avinash B S Krishnamurthy S, Lakshminarayanan V and Kumar S 2014 *Chem. Commun.* **50** 710.
[73] Avinash B S, Kumar M, Lakshminarayanan V and Kumar S 2015 *RSC Adv.* **5** 47692.

Acknowledgments
I would like to thank my students, colleagues and collaborators whose work is summarized in this paper.